MITCHELL/DELMAR ASE TEST COLLISION PREPARATION SERIES

Mitchell / Delmar Collision Test Preparation Handbook

Damage Analysis & Estimating

TEST B6

NOTICE TO THE READER

Publisher does not warrant or guarantee any of the products described herein or perform any independent analysis in connection with any of the product information contained herein. Publisher does not assume, and expressly disclaims, any obligation to obtain and include information other than that provided to it by the manufacturer.

The reader is expressly warned to consider and adopt all safety precautions that might be indicated by the activities herein and to avoid all potential hazards. By following the instructions contained herein, the reader willingly assumes all risks in connections with such instructions.

The publisher makes no representation or warranties of any kind, including but not limited to, the warranties of fitness for particular purpose or merchantability, nor are any such representations implied with respect to the material set forth herein, and the publisher takes no responsibility with respect to such material. The publisher shall not be liable for any special, consequential, or exemplary damages resulting, in whole or part, from the readers' use of, or reliance upon, this material.

Cover Design: Larry Barnett

Delmar Staff:
Publisher: Alar Elken
Acquisitions Editor: Vernon Anthony
Editorial Assistant: Betsy Hough
Production Editor: Dianne Jensis
Production Manager: Mary Ellen Black
Marketing Coordinator: Kasey Young
Marketing Manager: Mona Caron

Mitchell Staff:
Vice President and General Manager, Publications: Marc Brungger
Senior Product Manager: Judy Epstein
Vice President, Database Strategy & Licensing: Steve Hansen
Vice President, Database Communications & Development: Tom Fleming
Director, Parts Database, Data Acquisition: Ernie Tate
Director, Editorial Operations: Pat Rice
Manager, Database Marketing and Support: Deborah Arnold

Printed in Canada.

For more information contact:

Education, Government, and Retail Customers
Please contact Delmar at:
Phone: 1-800-347-7707
Fax: 1-606-647-5023
http://www.AutoEd.com

Business and Industry Customers
(Body shops, insurance companies, automotive repair facilities)
Please contact Mitchell at:
Phone: 1-800-238-9111, ext. 8508
Fax: 1-888-249-5942
http://www.mitchell.com

2 3 4 5 6 7 8 9 10 XXX 03 02 01 00 99

ISBN 0-7668-0571-9

Contents

Section 3 Are You Sure You're Ready for Test B6?

Section 4 An Overview of the System

Section 5 Sample Test for Practice

Section 6 Additional Test Questions for Practice

Section 7 Appendices

Preface

Mitchell International and Delmar Publishers have collaborated in publishing a brand new series of books designed to help technicians prepare for the ASE's collision repair tests. This collaboration unites Delmar, the leading publisher of automotive collision repair textbooks in North America with Mitchell International, the number-one source for collision repair and estimating information products for professionals. We are confident that this combined effort will bring you, the user of this book, the best test preparation tool possible.

With a total of five handbooks in the series, Mitchell and Delmar cover all of the ASE Collision Repair testing categories to help you prepare for and pass the following tests:

- Painting and Refinishing Test B2
- Non-Structural Analysis and Damage Repair Test B3
- Structural Analysis and Damage Repair Test B4
- Mechanical and Electrical Components Test B5
- Damage Analysis and Estimating Test B6

Each handbook in this series has been carefully researched and written to make sure you have all of the test preparation tools you need to succeed. Extensive interviews with many successful ASE test takers (both technicians and the shop owners who hire technicians) have provided us with the most wanted and most valuable information a test taker can have. What virtually everyone wanted were lots of practice tests and questions, and that's the first thing you'll notice in our series. Each handbook contains a general knowledge pretest, a sample test, and additional practice questions. No other test preparation guide gives you as much real-test practice time as Mitchell and Delmar's. We have worked hard to ensure that these questions match the ASE style in types of questions, quantities, and level of difficulty so you will be mentally prepared on test day. Also, the correct answers for all of these questions are provided.

Technicians also told us that they wanted to understand the ASE test and to have practical information about what they should expect. We have provided that as well, including a history of ASE and a section devoted to helping the technician "Take and Pass Every ASE Test" with case studies, test-taking strategies, and test formats.

Finally, technicians wanted refresher information and reference. Each of our books includes an overview section that is referenced to the task list. The complete task lists for each test appear in each book for the user's reference. There is also a complete glossary of terms for each book.

So whether you are looking for a sample test and a few extra questions to practice with or a complete introduction to ASE testing and support for preparing thoroughly, this *Mitchell and Delmar Handbook Series* is the best answer.

We believe you will benefit from this book, and we at Mitchell and Delmar wish you the best when it comes to passing your ASE tests. Thank you for choosing the *Mitchell and Delmar Handbook Series*. We welcome your comments and suggestions.

Alar Elken
Publisher, Technical Careers Team
Delmar Publishers

Marc Brungger
Vice President and General Manager
Mitchell Publications

The History of ASE

History

Originally known as The National Institute for Automotive Service Excellence (NIASE), today's ASE was founded in 1972 as a nonprofit, independent entity dedicated to improving the quality of automotive service and repair through the voluntary testing and certification of automotive technicians. Until that time, consumers had no way of distinguishing between competent and incompetent automotive technicians. In the mid-1960s and early 1970s, efforts were made by several automotive industry affiliated associations to respond to this need. Though the associations were nonprofit, many regarded certification test fees merely as a means of raising additional operating capital. Also, some associations, having a vested interest, produced test scores heavily weighted in the favor of its members.

NIASE

From these efforts a new independent, nonprofit association, the National Institute for Automotive Service Excellence (NIASE), was established much to the credit of two educators, George R. Kinsler, Director of Program Development for the Wisconsin Board of Vocational and Adult Education in Madison, WI, and Myron H. Appel, Division Chairman at Cypress College in Cypress, CA.

Early efforts were to encourage voluntary certification in four general areas:

TEST AREA	TITLES
I. Engine	Engines, Engine Tune-Up, Block Assembly, Cooling and Lube Systems, Induction, Ignition, and Exhaust
II. Transmission	Manual Transmissions, Drive Line and Rear Axles, and Automatic Transmissions
III. Brakes and Suspension	Brakes, Steering, Suspension, and Wheels
IV. Electrical/Air Conditioning	Body/Chassis, Electrical Systems, Heating, and Air Conditioning

In early NIASE tests, Mechanic A, Mechanic B type questions were used. Over the years the trend has not changed, but in mid-1984 the term was changed to Technician A, Technician B to better emphasize sophistication of the skills needed to perform successfully in the modern motor vehicle industry. In certain tests the term used is Estimator A/B, Painter A/B, or Parts Specialist A/B. At about that same time, the logo was changed from "The Gear" to "The Blue Seal," and the organization adopted the acronym ASE for Automotive Service Excellence.

Since those early beginnings, several other related trades have been added. ASE now administers a comprehensive series of certification exams for automotive and light truck repair technicians, medium and heavy truck repair technicians, alternate fuels technicians, engine machinists, collision repair technicians, school bus repair technicians, and parts specialists.

The Series and Individual Tests

- Automotive and Light Truck Technician; consisting of: Engine Repair—Automatic Transmission/Transaxle—Manual Drive Train and Axles—Suspension and Steering—Brakes—Electrical/Electronic Systems—Heating and Air Conditioning—Engine Performance
- Medium and Heavy Truck Technician; consisting of: Gasoline Engines—Diesel Engines—Drive Train—Brakes—Suspension and Steering—Electrical/Electronic Systems—HVAC—Preventive Maintenance Inspection
- Alternate Fuels Technician; consisting of: Compressed Natural Gas Light Vehicles
- Advanced Series; consisting of: Automobile Advanced Engine Performance and Advanced Diesel Engine Electronic Diesel Engine Specialty
- Collision Repair Technician; consisting of: Painting and Refinishing—Non-Structural Analysis and Damage Repair—Structural Analysis and Damage Repair—Mechanical and Electrical Components—Damage Analysis and Estimating
- Engine Machinist Technician; consisting of: Cylinder Head Specialist—Cylinder Block Specialist—Assembly Specialist
- School Bus Repair Technician; consisting of: Body Systems and Special Equipment—Drive Train—Brakes—Suspension and Steering—Electrical/Electronic Systems—Heating and Air Conditioning
- Parts Specialist; consisting of: Automobile Parts Specialist—Medium/Heavy Truck Parts Specialist

A Brief Chronology

1970–1971 Original questions were prepared by a group of forty auto mechanics teachers from public secondary schools, technical institutes, community colleges, and private vocational schools. These questions were then professionally edited by testing specialists at Educational Testing Service (ETS) at Princeton, New Jersey, and thoroughly reviewed by training specialists associated with domestic and import automotive companies.

1971 July: About eight hundred mechanics tried out the original test questions at experimental test administrations.

1972 November and December: Initial NIASE tests administered at 163 test centers. The original automotive test series consisted of four tests containing eighty questions each. Three hours were allotted for each test. Those who passed all four tests were designated Certified General Auto Mechanic (GAM).

1973 April and May: Test 4 was increased to 120 questions. Time was extended to four hours for this test. There were now 182 test centers. Shoulder patch insignias were made available.

	November: Automotive series expanded to five tests. Heavy-Duty Truck series of six tests introduced.
1974	November: Automatic Transmission (Light Repair) test modified. Name changed to Automatic Transmission.
1975	May: Collision Repair series of two tests is introduced.
1978	May: Automotive recertification testing is introduced.
1979	May: Heavy-Duty Truck recertification testing is introduced.
1980	May: Collision Repair recertification testing is introduced.
1982	May: Test administration providers switched from Educational Testing Service (ETS) to American College Testing (ACT). Name of Automobile Engine Tune-Up test changed to Engine Performance test.
1984	May: New logo was introduced. ASE's "The Blue Seal" replaced NIASE's "The Gear." All reference to Mechanic A, Mechanic B was changed to Technician A, Technician B.
1990	November: The first of the Engine Machinist test series was introduced.
1991	May: The second test of the Engine Machinist test series was introduced. November: The third and final Engine Machinist test was introduced.
1992	May: Name of Heavy-Duty Truck Test series changed to Medium/Heavy Truck test series.
1993	May: Automotive Parts Specialist test introduced. Collision Repair expanded to six tests. Light Vehicle Compressed Natural Gas test introduced. Limited testing begins in English-speaking provinces of Canada.
1994	May: Advanced Engine Performance Specialist test introduced.
1996	May: First three tests for School Bus Technician test series introduced. November: A Collision Repair test is added.
1997	May: A Medium/Heavy Truck test is added.
1998	May: A diesel advanced engine test is introduced: Electronic Diesel Engine Diagnosis Specialist. A test is added to the School Bus test series.

By the Numbers

Following are the approximate number of ASE technicians currently certified by category. The numbers may vary from time to time but are reasonably accurate for any given period. More accurate data may be obtained from ASE, which provides updates twice each year, in May and November after the Spring and Fall test series.

There are more than 338,000 Automotive Technicians with over 87,000 at Master Technician (MA) status. There are 47,000 Truck Technicians with over 19,000 at Master Technician (MT) status. There are 46,000 Collision Repair/Refinish Technicians with 7,300 at Master Technician (MB) status. There are 1,200 Estimators. There are 6,700 Engine Machinists with over 2,800 at Master Machinist Technician (MM) status. There are also 28,500 Automobile Advanced Engine Performance Technicians and over 2,700 School Bus Technicians for a combined total of more than 403,000 Repair Technicians. To this number, add over 22,000 Automobile Parts Specialists, and over 2,000 Truck Parts Specialists for a combined total of over 24,000 parts specialists.

There are over 6,400 ASE Technicians holding both Master Automotive Technician and Master Truck Technician status, of which 350 also hold Master Body Repair status. Almost 200 of these Master Technicians also hold Master Machinist status and five Technicians are certified in all ASE specialty areas.

Almost half of ASE certified technicians work in new vehicle dealerships (45.3 percent). The next greatest number work in independent garages with 19.8 percent. Next is tire dealerships with 9 percent, service stations at 6.3 percent, fleet shops at 5.7 percent, franchised volume retailers at 5.4 percent, paint and body shops at 4.3 percent, and specialty shops at 3.9 percent.

Of over 400,000 automotive technicians on ASE's certification rosters, almost 2,000 are female. The number of female technicians is increasing at a rate of about 20 percent each year. Women's increasing interest in the automotive industry is further evidenced by the fact that, according to the National Automobile Dealers Association (NADA), they influence 80 percent of the decisions of the purchase of a new automobile and represent 50 percent of all new car purchasers. Also, it is interesting to note that 65 percent of all repair and maintenance service customers are female.

The typical ASE certified technician is 36.5 years of age, is computer literate, deciphers a half-million pages of technical manuals, spends one hundred hours per year in training, holds four ASE certificates, and spends about $27,000 for tools and equipment. Twenty-seven percent of today's skilled ASE certified technicians attended college, many having earned an Associate of Science degree in Automotive Technology.

ASE

ASE's mission is to improve the quality of vehicle repair and service in the United States through the testing and certification of automotive repair technicians. Prospective candidates register for and take one or more of ASE's thirty-three exams. The tests are grouped into specialties for automobile, medium/heavy truck, school bus, and collision repair technicians as well as engine machinists, alternate fuels technicians, and parts specialists.

Upon passing at least one exam and providing proof of two years of related work experience, the technician becomes ASE certified. A technician who passes a series of exams earns ASE Master Technician status. An automobile technician, for example, must pass eight exams for this recognition.

The tests, conducted twice a year at over seven hundred locations around the country, are administered by American College Testing (ACT). They stress real-world diagnostic and repair problems. Though a good knowledge of theory is helpful to the technician in answering many of the questions, there are no questions specifically on theory. Certification is valid for five years. To retain certification, the technician must be retested to renew his or her certificate.

The automotive consumer benefits because ASE certification is a valuable yardstick by which to measure the knowledge and skills of individual technicians, as well as their commitment to their chosen profession. It is also a tribute to the repair facility employing ASE certified technicians. ASE certified technicians are permitted to wear blue and white ASE shoulder insignia, referred to as the "Blue Seal of Excellence," and carry credentials listing their areas of expertise. Often employers display their technicians' credentials in the customer waiting area. Customers look for facilities that display ASE's Blue Seal of Excellence logo on outdoor signs, in the customer waiting area, in the telephone book (Yellow Pages), and in newspaper advertisements.

The tests stress repair knowledge and skill. All test takers are issued a score report. In order to earn ASE certification, a technician must pass one or more of the exams and present proof of two years of relevant hands-on work experience. ASE certifications are valid for five years, after which time technicians must retest in order to keep up with changing technology and to remain in the ASE program. A nominal registration and test fee is charged.

To become part of the team that wears ASE's Blue Seal of Excellence®, please contact:

National Institute for Automotive Service Excellence
13505 Dulles Technology Drive
Herndon, VA 20171-3421

2 Take and Pass Every ASE Test

ASE Testing

Participating in an Automotive Service Excellence (ASE) voluntary certification program gives you a chance to show your customers that you have the "know-how" needed to work on today's modern vehicles. The ASE certification tests allow you to compare your skills and knowledge to the automotive service industry's standards for each specialty area.

If you are the "average" automotive technician taking this test, you are in your mid-thirties and have not attended school for about fifteen years. That means you probably have not taken a test in many years. Some of you, on the other hand, have attended college or taken postsecondary education courses and may be more familiar with taking tests and with test-taking strategies. There is, however, a difference in the ASE test you are preparing to take and the educational tests you may be accustomed to.

Who Writes the Questions?

The questions on an educational test are generally written, administered, and graded by an educator who may have little or no practical hands-on experience in the test area. The questions on all ASE tests are written by service industry experts familiar with all aspects of the subject area. ASE questions are entirely job-related and designed to test the skills that you need to know on the job.

The questions originate in an ASE "item-writing" workshop where service representatives from domestic and import automobile manufacturers, parts and equipment manufacturers, and vocational educators meet in a workshop setting to share their ideas and translate them into test questions. Each test question written by these experts is reviewed by all of the members of the group. The questions deal with the practical problems of diagnosis and repair that are experienced by technicians in their day-to-day hands-on work experiences.

All of the questions are pretested and quality-checked in a nonscoring section of tests by a national sample of certifying technicians. The questions that meet ASE's high standards of accuracy and quality are then included in the scoring sections of future tests. Those questions that do not pass ASE's stringent tests are sent back to the workshop or are discarded. ASE's tests are monitored by an independent proctor and are administered and machine-scored by an independent provider, American College Testing (ACT). All ASE tests have a three-year revision cycle.

Testing

If you think about it, we are actually tested on about everything we do. As infants, we were tested to see when we could turn over and crawl, later when we could walk or talk. As adolescents, we were tested to determine how well we learned the material presented in school and in how we demonstrated our accomplishments on the athletic field. As working adults, we are tested by our supervisors on how well we have completed an assignment or project. As nonworking adults, we are tested by our family on everyday activities, such as housekeeping or preparing a meal. Testing, then, is one of those facts of life that begins in the cradle and follows us to the grave.

Testing is an important fact of life that helps us to determine how well we have learned our trade. Also, tests often help us to determine what opportunities will be available to us in the future. To become ASE certified, we are required to take a test in every subject in which we wish to be recognized.

Be Test-Wise

In spite of the widespread use of tests, most technicians are not very test-wise. An ability to take tests and score well is a skill that must be acquired. Without this knowledge, the most intelligent and prepared technician may not do well on a test.

We will discuss some of the basic procedures necessary to follow in order to become a test-wise technician. Assume, if you will, that you have done the necessary study and preparation to score well on the ASE test.

Different approaches should be used for taking different types of tests. The different basic types of tests include: essay, objective, multiple-choice, fill-in-the-blank, true-false, problem solving, and open book. All ASE tests are of the four-part multiple-choice type.

Before discussing the multiple-choice type test questions, however, there are a few basic principles that should be followed before taking any test.

Before the Test

Do not arrive late. Always arrive well before your test is scheduled to begin. Allow ample time for the unexpected, such as traffic problems, so you will arrive on time and avoid the unnecessary anxiety of being late.

Always be certain to have plenty of supplies with you. For an ASE test, three or four sharpened soft lead (#2) pencils, a pocket pencil sharpener, erasers, and a watch are all that are required.

Do not listen to pretest chatter. When you arrive early, you may hear other technicians testing each other on various topics or making their best guess as to the probable test questions. At this time, it is too late to add to your knowledge. Also the rhetoric may only confuse you. If you find it bothersome, take a walk outside the test room to relax and loosen up.

Read and listen to all instructions. It is important to read and listen to the instructions. Make certain that you know what is expected of you. Listen carefully to verbal instructions and pay particular attention to any written instructions on the test paper. Do not dive into answering questions only to find out that you have answered the wrong question by not following instructions carefully. It is difficult to make a high score on a test if you answer the wrong questions.

These basic principles have been violated in most every test ever given. Try to remember them. They are essential for success.

Objective Tests

A test is called an objective test if the same standards and conditions apply to everyone taking the test and there is only one correct answer to each question. Objective tests primarily measure your ability to recall information. A well-designed objective test can also test your ability to understand, analyze, interpret, and apply your knowledge. Objective tests include true-false, multiple-choice, fill-in-the-blank, and matching questions.

Objective questions, not generally encountered in a classroom setting, are frequently used in standardized examinations. Objective tests are easy to grade and also reduce the amount of paperwork necessary to administer. The objective tests are used in entry-level programs or when very large numbers are being tested. ASE's tests consist exclusively of four-part multiple-choice objective questions in all of their tests.

Taking an Objective Test

The principles of taking an objective test are somewhat different from those used in other types of tests. You should first quickly look over the test to determine the number of questions, but do not try to read through all of the questions. In an ASE test, there are usually between forty and eighty questions, depending on the subject matter. Read through each question before marking your answer. Answer the questions in the order they appear on the test. Leave the questions blank that you are not sure of and move on to the next question. You can return to those unanswered questions after you have finished the others. They may be easier to answer at a later time after your mind has had additional time to consider them on a subconscious level. In addition, you might find information in other questions that will help you to answer some of them.

Do not be obsessed by the apparent pattern of responses. For example, do not be influenced by a pattern like **d**, **c**, **b**, **a**, **d**, **c**, **b**, **a** on an ASE test.

There is also a lot of folk wisdom about taking objective tests. For example, there are those who would advise you to avoid response options that use certain words such as *all*, *none*, *always*, *never*, *must*, and *only*, to name a few. This, they claim, is because nothing in life is exclusive. They would advise you to choose response options that use words that allow for some exception, such as *sometimes*, *frequently*, *rarely*, *often*, *usually*, *seldom*, and *normally*. They would also advise you to avoid the first and last option (A and D) because test writers, they feel, are more comfortable if they put the correct answer in the middle (B and C) of the choices. Another recommendation often offered is to select the option that is either shorter or longer than the other three choices because it is more likely to be correct. Some would advise you to never change an answer since your first intuition is usually correct.

Although there may be a grain of truth in this folk wisdom, ASE test writers try to avoid them and so should you. There are just as many **A** answers as there are **B** answers, just as many **D** answers as **C** answers. As a matter of fact, ASE tries to balance the answers at about 25 percent per choice **A**, **B**, **C**, and **D**. There is no intention to use "tricky" words, such as outlined above. Put no credence in the opposing words "sometimes" and "never," for example. When used in an ASE type question, one or both may be correct; one or both may be incorrect.

There are some special principles to observe on multiple-choice tests. These tests are sometimes challenging because there are often several choices that may seem possible, and it may be difficult to decide on the correct choice. The best strategy, in this case, is to first determine the correct answer before looking at the options. If you see the answer you decided on, you should still examine the options to make sure that none seem more correct than yours. If you do not know or are not sure of the answer, read each option very carefully and try to eliminate those options that you know to be wrong. That way, you can often arrive at the correct choice through a process of elimination.

If you have gone through all of the test and you still do not know the answer to some of the questions, then guess. Yes, guess. You then have at least a 25 percent chance of being correct. If you leave the question blank, you have no chance. In ASE tests, there is no penalty for being wrong. As the late President Franklin D. Roosevelt once advised a group of students, "It is common sense to take a method and try it. If it fails, admit it frankly and try another. But above all, try something."

During the Test

Mark your bubble sheet clearly and accurately. One of the biggest problems an adult faces in test-taking, it seems, is in placing an answer in the correct spot on a bubble sheet. Make certain that you mark your answer for, say, question 21, in the space on the bubble sheet designated for the answer for question 21. A correct response in the wrong bubble will probably be wrong. Remember, the answer sheet is machine scored and can only "read" what you have bubbled in. Also, do not bubble in two answers for the same question. For example, if you feel the answer to a particular question is **A** but think it may be **C,** do not bubble in both choices. Even if either **A** or **C** is correct, a double answer will score as an incorrect answer. It's better to take a chance with your best guess.

Review Your Answers

If you finish answering all of the questions on a test ahead of time, go back and review the answers of those questions that you were not sure of. You can often catch careless errors by using the remaining time to review your answers.

Don't Be Distracted

At practically every test, some technicians will invariably finish ahead of time and turn their papers in long before the final call. Do not let them distract or intimidate you. Either they knew too little and could not finish the test, or they were very self-confident and thought they knew it all. Perhaps they were trying to impress the proctor or other technicians about how much they know. Often you may hear them later talking about the information they knew all the while but forgot to respond on their answer sheet.

Use Your Time Wisely

It is not wise to use less than the total amount of time that you are allotted for a test. If there are any doubts, take the time for review. Any product can usually be made better with some additional effort. A test is no exception. It is not necessary to turn in your test paper until you are told to do so.

Don't Cheat

Some technicians may try to use a "crib sheet" during a test. Others may attempt to read answers from another technician's paper. If you do that, you are unquestionably assuming that someone else has a correct answer. You probably know as much, maybe more, than anyone else in the test room. Trust yourself. If you're still not convinced, think of the consequences of being caught. Cheating is foolish. If you are caught, you have failed the test.

Be Confident

The first and foremost principle in taking a test is that you need to know what you are doing, to be test-wise. It will now be presumed that you are a test-wise technician and are now ready for some of the more obscure aspects of test-taking.

An ASE-style test requires that you use the information and knowledge at your command to solve a problem. This generally requires a combination of information similar to the way you approach problems in the real world. Most problems, it seems, typically do not fall into neat textbook cases. New problems are often difficult to handle, whether they are encountered inside or outside the test room.

An ASE test also requires that you apply methods taught in class as well as those learned on the job to solve problems. These methods are akin to a well-equipped tool box in the hands of a skilled technician. You have to know what tools to use in a particular situation, and you must also know how to use them. In an ASE test, you will need to be able to demonstrate that you are familiar with and know how to use the tools.

You should begin a test with a completely open mind. At times, however, you may have to move out of your normal way of thinking and be creative to arrive at a correct answer. If you have diligently studied for at least one week prior to the test, you have bombarded your mind with a wide assortment of information. Your mind will be working with this information on a subconscious level, exploring the interrelationships among various facts, principles, and ideas. This prior preparation should put you in a creative mood for the test.

In order to reach your full potential, you should begin a test with the proper mental attitude and a high degree of self-confidence. You should think of a test as an opportunity to document how much you know about the various tasks in your chosen profession. If you have been diligently studying the subject matter, you will be able to take your test in serenity because your mind will be well organized. If you are confident, you are more likely to do well because you have the proper mental attitude. If, on the other hand, your confidence is low, you are bound to do poorly. It is a self-fulfilling prophecy.

Perhaps you have heard athletic coaches talk about the importance of confidence when competing in sports. Mental confidence helps an athlete to perform at the highest level and gain an advantage over competitors. Taking a test is much like an athletic event. You are competing against yourself, in a certain sense, because you will be trying to approach perfection in determining your answers. As in any competition, you should aim your sights high and be confident that you can reach the apex.

Anxiety and Fear

Many technicians experience anxiety and fear at the very thought of taking a test. Many worry, become nervous, and even become ill at test time because of the fear of failure. Many often worry about the criticism and ridicule that may come from their employer, relatives, and peers. Some worry about taking a test because they feel that the stakes are very high. Those who spent a great amount of time studying may feel they must get a high grade to justify their efforts. The thought of not doing well can result in unnecessary worry. They become so worried, in fact, that their reasoning and thinking ability is impaired, actually bringing about the problem they wanted to avoid.

The fear of failure should not be confused with the desire for success. It is natural to become "psyched-up" for a test in contemplation of what is to come. A little emotion can provide a healthy flow of adrenaline to peak your senses and hone your mental ability. This improves your performance on the test and is a very different reaction from fear.

Most technician's fears and insecurities experienced before a test are due to a lack of self-confidence. Those who have not scored well on previous tests or have no confidence in their preparation are those most likely to fail. Be confident that you will do well on your test and your fears should vanish. You will know that you have done everything possible to realize your potential.

Getting Rid of Fear

If you have previously experienced fear of taking a test, it may be difficult to change your attitude immediately. It may be easier to cope with fear if you have a better understanding of what the test is about. A test is merely an assessment of how much the technician knows about a particular task area. Tests, then, are much less threatening when thought of in this manner. This does not mean, however, that you should lower your self-esteem simply because you performed poorly on a test.

You can consider the test essentially as a learning device, providing you with valuable information to evaluate your performance and knowledge. Recognize that no one is perfect. All humans make mistakes. The idea, then, is to make mistakes before the test, learn from them, and avoid repeating them on the test. Fortunately, this is not as difficult as it seems. Practical questions in this study guide include the correct answers to consider if you have made mistakes on the practice test. You should learn where you went wrong so you will not repeat them in the ASE test. If you learn from your mistakes, the stage is set for future growth.

If you understood everything presented up until now, you have the knowledge to become a test-wise technician, but more is required. To be a test-wise technician, you not only have to practice these principles, you have to diligently study in your task area.

Effective Study

The fundamental and vital requirement to induce effective study is a genuine and intense desire to achieve. This is more basic than any rule or technique that will be given here. The key requirement, then, is a driving motivation to learn and to achieve.

If you wish to study effectively, first develop a desire to master your studies and sincerely believe that you will master them. Everything else is secondary to such a desire.

First, build up definite ambitions and ideals toward which your studies can lead. Picture the satisfaction of success. The attitude of the technician may be transformed from merely getting by to an earnest and energetic effort. The best direct stimulus to change may involve nothing more than the deliberate planning of your time. Plan time to study.

Another drive that creates positive study is an interest in the subject studied. As an automotive technician, you can develop an interest in studying particular subjects if you follow these four rules:

1. Acquire information from a variety of sources. The greater your interest in a subject, the easier it is to learn about it. Visit your local library and seek books on the subject you are studying. When you find something new or of interest, make inexpensive photocopies for future study.
2. Merge new information with your previous knowledge. Discover the relationship of new facts to old known facts. Modern developments in automotive technology take on new interest when they are seen in relation to present knowledge.
3. Make new information personal. Relate the new information to matters that are of concern to you. The information you are now reading, for example, has interest to you as you think about how it can help.
4. Use your new knowledge. Raise questions about the points made by the book. Try to anticipate what the next steps and conclusions will be. Discuss this new knowledge, particularly the difficult and questionable points, with your peers.

You will find that when you study with eager interest, you will discover it is no longer work. It is pleasure and you will be fascinated in what you study. Studying can be like reading a novel or seeing a movie that overcomes distractions and requires no effort or willpower. You will discover that the positive relationship between interest and effort works both ways. Even though you perhaps began your studies with little or no interest, simply staying with it helped you to develop an interest in your studies.

Obviously, certain subject matter studies are bound to be of little or no interest, particularly in the beginning. Parts of certain studies may continue to be uninteresting. An honest effort to master those subjects, however, nearly always brings about some level of interest. If you appreciate the necessity and reward of effective studying, you will rarely be disappointed. Here are a few important hints for gaining the determination that is essential to carrying good conclusions into actual practice.

Make Study Definite

Decide what is to be studied and when it is to be studied. If the unit is discouragingly long, break it into two or more parts. Determine exactly what is involved in the first part and learn that. Only then should you proceed to the next part. Stick to a schedule.

The Urge to Learn

Make clear to yourself the relation of your present knowledge to your study materials. Determine the relevance with regard to your long-range goals and ambitions.

Turn your attention away from real or imagined difficulties as well as other things that you would rather be doing. Some major distractions are thoughts of other duties and of disturbing problems. These distractions can usually be put aside, simply shunted off by listing them in a notebook. Most technicians have found that by writing interfering thoughts down, their minds are freed from annoying tensions.

Adopt the most reasonable solution you can find or seek objective help from someone else for personal problems. Personal problems and worry are often causes of ineffective study. Sometimes there are no satisfactory solutions. Some manage to avoid the problems or to meet them without great worry. For those who may wish to find better ways of meeting their personal problems, the following suggestions are offered:

1. Determine as objectively and as definitely as possible where the problem lies. What changes are needed to remove the problem, and which changes, if any, can be made? Sometimes it is wiser to alter your goals than external conditions. If there is no perfect solution, explore the others. Some solutions may be better than others.
2. Seek an understanding confidant who may be able to help analyze and meet your problems. Very often, talking over your problems with someone in whom you have confidence and trust will help you to arrive at a solution.
3. Do not betray yourself by trying to evade the problem or by pretending that it has been solved. If social problem distractions prevent you from studying or doing satisfactory work, it is better to admit this to yourself. You can then decide what can be done about it.

Once you are free of interferences and irritations, it is much easier to stay focused on your studies.

Concentrate

To study effectively, you must concentrate. Your ability to concentrate is governed, to a great extent, by your surroundings as well as your physical condition. When absorbed in study, you must be oblivious to everything else around you. As you learn to concentrate and study, you must also learn to overcome all distractions. There are three kinds of distractions you may face:

1. Distractions in the surrounding area, such as motion, noise, and the glare of lights. The sun shining through a window on your study area, for example, can be very distracting.

   Some technicians find that, for effective study, it is necessary to eliminate visual distractions as well as noises. Others find that they are able to tolerate moderate levels of auditory or visual distraction.

   Make sure your study area is properly lighted and ventilated. The lighting should be adequate but should not shine directly into your eyes or be visible out of the corner of your eye. Also, try to avoid a reflection of the lighting on the pages of your book.

   Whether heated or cooled, the environment should be at a comfortable level. For most, this means a temperature of 78°F–80°F (25.6°C–26.7°C) with a relative humidity of 45 to 50 percent.

2. Distractions arising from your body, such as a headache, fatigue, and hunger. Be in good physical condition. Eat wholesome meals at regular times. Try to eat with your family or friends whenever possible. Mealtime should be your recreational period. Do not eat a heavy meal for lunch, and do not resume studies immediately after eating lunch. Just after lunch, try to get some regular exercise, relaxation, and recreation. A little exercise on a regular basis is much more valuable than a lot of exercise only on occasion.
3. Distractions of irrelevant ideas, such as how to repair the garden gate, when you are studying for an automotive-related test.

The problems associated with study are no small matter. These problems of distractions are generally best dealt with by a process of elimination. A few important rules for eliminating distractions follow.

Get Sufficient Sleep

You must get plenty of rest even if it means dropping certain outside activities. Avoid cutting in on your sleep time; you will be rewarded in the long run. If you experience difficulty going to sleep, do something to take your mind off your work and try to relax before going to bed. Some suggestions that may help include a little reading, a warm bath, a short walk, a conversation with a friend, or writing that overdue letter to a distant relative. If sleeplessness is an ongoing problem, consult a physician. Do not try any of the sleep remedies on the market, particularly if you are on medication, without approval of your physician.

If you still have difficulty studying, a final rule may help. Sit down in a favorable place for studying, open your books, and take out your pencil and paper. In a word, go through the motions.

Arrange Your Area

Arrange your chair and work area. To avoid strain and fatigue, whenever possible, shift your position occasionally. Try to be comfortable; however, avoid being too comfortable. It is nearly impossible to study rigorously when settled back in a large easy chair or reclining leisurely on a sofa.

When studying, it is essential to have a plan of action, a time to work, a time to study, and a time for pleasure. If you schedule your day and adhere to the schedule, you will eliminate most of your efforts and worries. A plan that is followed, then, soon becomes the easy and natural routine of the day. Most technicians find it useful to have a definite place and time to study. A particular table and chair should always be used for study and intellectual work. This place will then come to mean study. To be seated in that particular location at a regularly scheduled time will automatically lead you to assume a readiness for study.

Don't Daydream

Daydreaming or mind-wandering is an enemy of effective study. Daydreaming is frequently due to an inadequate understanding of words. Use the Glossary or a dictionary to look up the troublesome word. Another frequent cause of daydreaming is a deficient background in the present subject matter. When this is the problem, go back and review the subject matter to obtain the necessary foundation. Just one hour of concentrated study is equivalent to ten hours with frequent lapses of daydreaming. Be on guard against mind-wandering, and pull yourself back into focus on every occasion.

Study Regularly

A system of regularity in study is believed by many scholars to be the secret of success. The daily time schedule must, however, be determined on an individual basis. You must decide how many hours of each day you can devote to your studies. Few technicians really are aware of where their leisure time is spent. An accurate account of how your days are presently being spent is an important first step toward creating an effective daily schedule.

Weekly Schedule							
	Sun	Mon	Tues	Wed	Thu	Fri	Sat
6:00							
6:30							
7:00							
7:30							
8:00							
8:30							
9:00							
9:30							
10:00							
10:30							
11:00							
11:30							
NOON							
12:30							
1:00							
1:30							
2:00							
2:30							
3:00							
3:30							
4:00							
4:30							
5:00							
5:30							
6:00							
6:30							
7:00							
7:30							
8:00							
8:30							
9:00							
9:30							
10:00							
10:30							
11:00							
11:30							

The convenient form is for keeping an hourly record of your week's activities. If you fill in the schedule each evening before bedtime, you will soon gain some interesting and useful facts about yourself and your use of your time. If you think over the causes of wasted time, you can determine how you might better spend your time. A practical schedule can be set up by using the following steps.

1. Mark your fixed commitments, such as work, on your schedule. Be sure to include classes and clubs. Do you have sufficient time left? You can arrive at an estimate of the time you need for studying by counting the hours used during the present week. An often-used formula, if you are taking classes, is to multiply the number of hours you spend in class by two. This provides time for class studies. This is then added to your work hours. Do not forget time allocation for travel.
2. Fill in your schedule for meals and studying. Use as much time as you have available during the normal workday hours. Do not plan, for example, to do all of your studying between 11:00 P.M. and 1:00 A.M. Try to select a time for study that you can use every day without interruption. You may have to use two or perhaps three different study periods during the day.
3. List the things you need to do within a time period. A one-week time frame seems to work well for most technicians. The question you may ask yourself is: "What do I need to do to be able to walk into the test next week, or next month, prepared to pass?"
4. Break down each task into smaller tasks. The amount of time given to each area must also be settled. In what order will you tackle your schedule? It is best to plan the approximate time for your assignments and the order in which you will do them. In this way, you can avoid the difficulties of not knowing what to do first and of worrying about the other things you should be doing.
5. List your tasks in the empty spaces on your schedule. Keep some free time unscheduled so you can deal with any unexpected events, such as a dental appointment. You will then have a tentative schedule for the following week. It should be flexible enough to allow some units to be rearranged if necessary. Your schedule should allow time off from your studies. Some use the promise of a planned recreational period as a reward for motivating faithfulness to a schedule. You will more likely lose control of your schedule if it is packed too tightly.

Keep a Record

Keep a record of what you actually do. Use the knowledge you gain by keeping a record of what you are actually doing so you can create or modify a schedule for the following week. Be sure to give yourself credit for movement toward your goals and objectives. If you find that you cannot study productively at a particular hour, modify your schedule so as to correct that problem.

Scoring the ASE Test

You can gain a better perspective about tests if you know and understand how they are scored. ASE's tests are scored by American College Testing (ACT), a nonpartial, nonbiased organization having no vested interest in ASE or in the automotive industry. Each question carries the same weight as any other question. For example, if there are fifty questions, each is worth 2 percent of the total score. The passing grade is 70 percent. That means you must correctly answer thirty-five of the fifty questions to pass the test.

Understand the Test Results

The test results can tell you:

- where your knowledge equals or exceeds that needed for competent performance, or
- where you might need more preparation.

The test results *cannot* tell you:

- how you compare with other technicians, or
- how many questions you answered correctly.

Your ASE test score report will show the number of correct answers you got in each of the content areas. These numbers provide information about your performance in each area of the test. However, because there may be a different number of questions in each area of the test, a high percentage of correct answers in an area with few questions may not offset a low percentage in an area with many questions.

It may be noted that one does not "fail" an ASE test. The technician that does not pass is simply told "More Preparation Needed." Though large differences in percentages may indicate problem areas, it is important to consider how many questions were asked in each area. Since each test evaluates all phases of the work involved in a service specialty, you should be prepared in each area. A low score in one area could keep you from passing an entire test.

Note that a typical test will contain the number of questions indicated above each content area's description. For example:

Damage Analysis and Estimating (Test B6)

Content Area	Questions	Percent of Test
A. Damage Analysis	11	22%
B. Estimating	14	28%
C. Legal and Environmental Practices	2	4%
D. Vehicle Construction	6	12%
E. Vehicle Systems Knowledge	11	22%
1. Fuel, Intake, Ignition, and Exhaust Systems (1)		
2. Suspension, Steering, and Powertrain (3)		
3. Brakes (1)		
4. Heating, Engine Cooling, and Air Conditioning (2)		
5. Electrical/Electronic Systems (1)		
6. Safety Systems (2)		
7. Fasteners and Materials (1)		
F. Parts Identification and Source Determination	4	8%
1. New Original Equipment Manufacturer (OEM) (1)		
2. New Aftermarket (1)		
3. Salvage/Used (1)		
4. Remanufactured/Rebuilt/Reconditioned (1)		
G. Customer Relations and Sales Skills	2	4%
TOTAL	*50	100%

***Note:** The test could contain up to ten additional questions that are included for statistical research purposes only. Your answers to these questions will not affect your score, but since you do not know which ones they are, you should answer all questions in the test. The five-year Recertification Test will cover the same content areas as those listed above. However, the number of questions in each content area of the Recertification Test will be reduced by about one-half.*

"Average"

There is no such thing as average. You cannot determine your overall test score by adding the percentages given for each task area and dividing by the number of areas. It doesn't work that way because there generally are not the same number of questions in each task area. A task area with twenty questions, for example, counts more toward your total score than a task area with ten questions.

So, How Did You Do?

Your test report should give you a good picture of your results and a better understanding of your task areas of strength and weakness.

If you fail to pass the test, you may take it again at any time it is scheduled to be administered. You are the only one who will receive your test score. Test scores will not be given over the telephone by ASE nor will they be released to anyone without your written permission.

3 Are You Sure You're Ready for Test B6?

Pretest

The purpose of this pretest is to determine the amount of review that you may require prior to taking the ASE collision repair/refinish test: Damage Analysis and Estimating (Test B6). If you answer all of the pretest questions correctly, complete the sample test in section 5 along with the additional test questions in section 6.

If two or more of your answers to the pretest questions are wrong, study section 4: An Overview of the System before continuing with the sample test and additional test questions.

The pretest answers and explanations are located at the end of the pretest.

1. Which of the following can be considered a structural part of the unibody vehicle?
   A. Door intrusion beam
   B. Back window glass
   C. Bumper reinforcement
   D. Engine mount
2. ACV means
   A. A crashed vehicle
   B. Actual cash value
   C. Approximate car value
   D. Approximate collision value
3. Estimator A says that the first step in a four wheel alignment is thrust angle. Estimator B says that castor is the inward or outward tilt of a wheel as viewed from the front. Who is right?
   A. A only
   B. B only
   C. Both A and B
   D. Neither A nor B
4. Estimator A says that when estimating collision damage, you should never replace a sheet metal part with anything other than an OEM part. Estimator B says that you should use OEM parts for safety-related items. Who is right?
   A. A only
   B. B only
   C. Both A and B
   D. Neither A nor B

5. All of the following are unibody structural parts **EXCEPT:**
   A. lower frame rail.
   B. rocker panel.
   C. fender.
   D. radiator support.
6. All of the following are included in the VIN **EXCEPT:**
   A. paint code.
   B. model year.
   C. engine type.
   D. sequential production number.

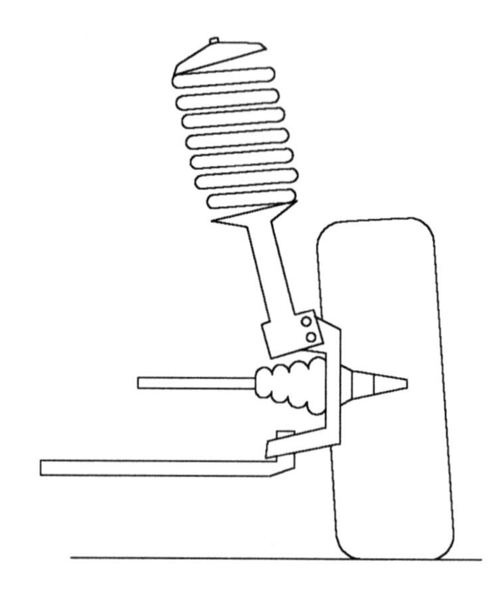

7. While writing an estimate, the estimator notices that the left front tire is tilted as shown above. The estimator should:
   A. make a note of the damage in the estimate.
   B. check for bent components.
   C. add in an alignment.
   D. align the vehicle before repairs.

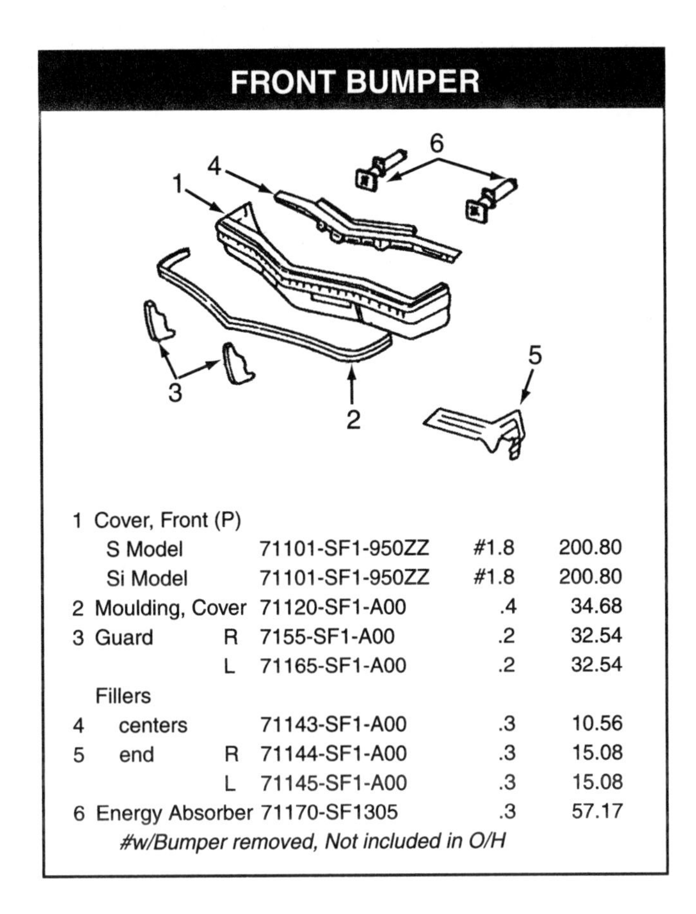

1 Cover, Front (P)				
S Model		71101-SF1-950ZZ	#1.8	200.80
Si Model		71101-SF1-950ZZ	#1.8	200.80
2 Moulding, Cover		71120-SF1-A00	.4	34.68
3 Guard	R	7155-SF1-A00	.2	32.54
	L	71165-SF1-A00	.2	32.54
Fillers				
4 centers		71143-SF1-A00	.3	10.56
5 end	R	71144-SF1-A00	.3	15.08
	L	71145-SF1-A00	.3	15.08
6 Energy Absorber		71170-SF1305	.3	57.17

#w/Bumper removed, Not included in O/H

8. Estimator A writes an estimate that shows the paint time of the right and left end fillers as 0.5 each. Estimator B writes an estimate stating that the R&R of the energy absorber is 0.3 each, after bumper removal. Using the above figure, who is right?
   A. A only
   B. B only
   C. Both A and B
   D. Neither A nor B
9. What part is most likely damaged in a head-on collision between two unibody vehicles?
   A. Front door
   B. Cowl
   C. Lower frame rail
   D. Shock tower
10. A new, less than one-year-old vehicle is damaged. What type of parts will most likely be installed?
    A. Salvage
    B. Aftermarket
    C. OEM new
    D. Rebuilt

11. The left front wheel of a unibody with a McPherson strut suspension is tilted in, when viewed from the front. Damage to which part is least likely to be the cause:
    A. strut.
    B. steering knuckle.
    C. control arm.
    D. stabilizer bar.
12. A ten-year old vehicle is damaged. What type of parts are least likely to be installed?
    A. Salvage
    B. Aftermarket
    C. OEM new
    D. Rebuilt

Answers to the Test Questions for the Pretest

1. B, 2. B, 3. A, 4. B, 5. C, 6. A, 7. B, 8. B, 9. C, 10. C, 11. D, 12. C

Explanations to the Answers for the Pretest

Question #1
Answer A is wrong. Door intrusion beams are not structural parts.
Answer B is correct. Urethane set back glass adds rigidity to the rear window opening and is considered a structural part.
Answer C is wrong. Bumper reinforcements are not structural parts.
Answer D is wrong. An engine mount is not a structural part.

Question #2
Answer A is wrong.
Answer B is correct. ACV means actual cash value.
Answer C is wrong.
Answer D is wrong.

Question #3
Answer A is correct. The thrust angle must be correct before the front wheels can be set.
Answer B is wrong. Camber is the tilt of the top of the wheel, viewed from the front.
Answer C is wrong.
Answer D is wrong.

Question #4
Answer A is wrong. Sheet metal can be replaced with aftermarket parts.
Answer B is correct. Safety components should be replaced with OEM.
Answer C is wrong.
Answer D is wrong.

Question #5
Answer A is wrong. A lower frame rail is a structural part.
Answer B is wrong. A rocker panel is a structural part.
Answer C is correct. A fender is not a structural part.
Answer D is wrong. A radiator support is a structural part.

Question #6
Answer A is correct. The paint code is not part of the VIN.
Answer B is wrong. The model year is part of the VIN.
Answer C is wrong. The engine type is part of the VIN.
Answer D is wrong. The sequential production number is part of the VIN.

Question #7
Answer A is wrong. The damaged parts should be listed on the estimate.
Answer B is correct. The suspension should be checked for damage.
Answer C is wrong. An alignment will be needed but the damaged parts must be identified.
Answer D is wrong. An alignment will be needed after the repairs are made.

Question #8
Answer A is wrong. There is only replacement time listed, not paint time.
Answer B is correct. The R&R time of the absorbers is 0.3 each, with the bumpers removed.
Answer C is wrong.
Answer D is wrong.

Question #9
Answer A is wrong. The front door is not likely damaged in this situation.
Answer B is wrong. The cowl may be damaged, but it is not the most likely.
Answer C is correct. The lower frame rail is the most likely to be damaged.
Answer D is wrong. The shock tower may be damaged, but it is not the most likely.

Question #10
Answer A is wrong.
Answer B is wrong.
Answer C is correct. A new vehicle would most likely have OEM new parts installed.
Answer D is wrong.

Question #11
Answer A is wrong. A bent strut may cause a camber problem.
Answer B is wrong. A bent steering knuckle may cause a camber problem.
Answer C is wrong. A bent control arm may cause a camber problem.
Answer D is correct. A bent stabilizer bar is the least likely to cause a camber problem.

Question #12
Answer A is wrong.
Answer B is wrong.
Answer C is correct. OEM new parts are the least likely to be installed on a ten-year old vehicle.
Answer D is wrong.

Types of Questions

ASE certification tests are often thought of as being tricky. They may seem to be tricky if you do not completely understand what is being asked. The following examples will help you recognize certain types of ASE questions and avoid common errors.

Each test is made up of forty to eighty multiple-choice questions. Multiple-choice questions are an efficient way to test knowledge. To answer them correctly, you must think about each choice as a possibility, and then choose the one that best answers the question. To do this, read each word of the question carefully. Do not assume you know what the question is about until you have finished reading it.

Multiple-Choice Questions

One type of multiple-choice question has three wrong answers and one correct answer. The wrong answers, however, may be almost correct, so be careful not to jump at the first answer that seems to be correct. If all the answers seem to be correct, choose the answer that is the most correct. If you readily know the answer, this kind of question does not present a problem. If you are unsure of the answer, analyze the question and the answers. For example:

Question 1:

The most common type of steel on a vehicle is:

A. ultra high-strength steel.
B. mild steel.
C. high strength steel.
D. martensetic steel.

Analysis:

Answer A is wrong. Ultra high strength steel is not the most common type of steel on a vehicle.
Answer B is correct. Mild steel is the most common type of steel on a vehicle.
Answer C is wrong. High strength steel is not most common type of steel on a vehicle.
Answer D is wrong. Martensetic steel is a type of ultra high strength steel.

EXCEPT Questions

Another type of question used on ASE tests has answers that are all correct except one. The correct answer for this type of question is the answer that is wrong. The word "EXCEPT" will always be in capital letters. You must identify which of the choices is the wrong answer. If you read quickly through the question, you may overlook what the question is asking and answer the question with the first correct statement. This will make your answer wrong. An example of this type of question and the analysis is as follows:

Question 2:

All of the following can be used to measure the center line **EXCEPT:**

A. laser.
B. tram gauge.
C. universal bench.
D. self-centering gauges.

Analysis:
Answer A is wrong. A laser can be used to measure center line
Answer B is correct. A tram gauge can not be used to measure center line
Answer C is wrong. A universal bench can be used to measure center line
Answer D is wrong. Self centering gauge can be used to measure center line

Estimator A, Estimator B Questions

The type of question that is most popularly associated with an ASE test is the "Estimator A says... Estimator B says... Who is right?" type. In this type of question, you must identify the correct statement or statements. To answer this type of question correctly, you must carefully read each estimator's statement and judge it on its own merit to determine if the statement is true.

Typically, this type of question begins with a statement about some analysis or repair procedure. This is followed by two statements about the cause of the problem, proper inspection, identification, or repair choices. You are asked whether the first statement, the second statement, both statements, or neither statement is correct. Analyzing this type of question is a little easier than the other types because there are only two ideas to consider although there are still four choices for an answer.

Technician A... Technician B questions are really double-true-false questions. The best way to analyze this kind of question is to consider each estimator's statement separately. Ask yourself, is A true or false? Is B true or false? Then select your answer from the four choices. An important point to remember is that an ASE Technician A... Technician B question will never have Technician A and B directly disagreeing with each other. That is why you must evaluate each statement independently. An example of this type of question and the analysis of it follows.

Question 3:

Estimator A says that slightly damaged suspension components can be straightened. Estimator B says that the estimator must determine the most cost effective method to restore the damaged vehicle to pre-accident condition. Who is right?

A. A only
B. B only
C. Both A and B
D. Neither A nor B

Analysis:
Answer A is wrong. Damaged suspension components must be replaced.
Answer B is correct. The most cost effective repair method must be determined.
Answer C is wrong.
Answer D is wrong.

Questions with a Figure

About 10 percent of ASE questions will have a figure, as shown in the following example:

Question 4:

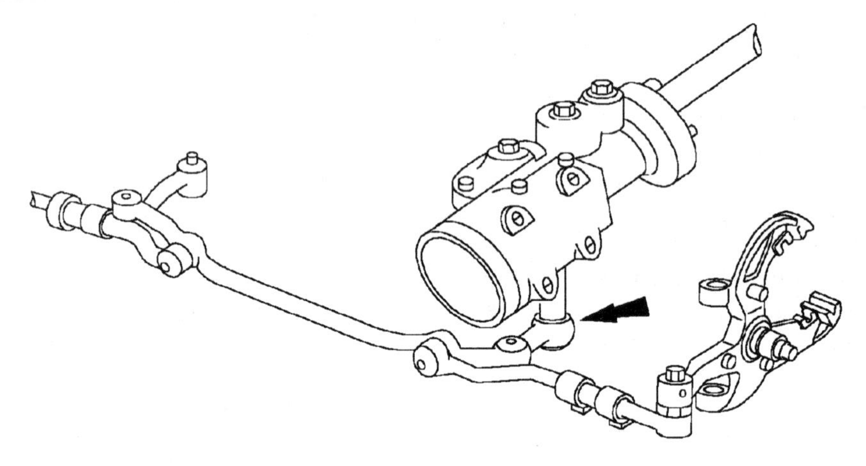

The component shown above is:

A. a connector link.
B. an idler arm.
C. a pitman arm.
D. a connecting rod.

Analysis:

Answer A is wrong.
Answer B is wrong.
Answer C is correct. The arrow indicates the pitman arm.
Answer D is wrong.

Most-Likely Questions

Most-likely questions are somewhat difficult because only one choice is correct while the other three choices are nearly correct. An example of a most-likely-cause question is as follows:

Question 5:

Which of the following parts would be most likely to be available aftermarket:

A. quarter panel.
B. apron.
C. center pillar.
D. fender.

Analysis:

Answer A is wrong.
Answer B is wrong.
Answer C is wrong.
Answer D is correct. The fender is the most likely of these parts to be available as aftermarket.

Least-Likely Questions

Notice that in most-likely questions there is no capitalization. This is not so with least-likely type questions. For this type of question, look for the choice that would be the least likely cause of the described situation. Read the entire question carefully before choosing your answer. An example is as follows:

Question 6:

. Which of the following electrical components is LEAST likely to be damaged in a frontal collision?

A. Head light
B. Starter
C. Turn signal lamp
D. Battery

Analysis:

Answer A is wrong. The head light is likely damaged in a frontal collision.
Answer B is correct. The starter is the least likely of the listed items to be damaged.
Answer C is wrong. The turn signal lamp is likely damaged in a frontal collision.
Answer D is wrong. The battery is more likely to be damaged than the starter.

Summary

There are no four-part multiple-choice ASE questions having "none of the above" or "all of the above" choices. ASE does not use other types of questions, such as fill-in-the-blank, completion, true-false, word-matching, or essay. ASE does not require you to draw diagrams or sketches. If a formula or chart is required to answer a question, it is provided for you. There are no ASE questions that require you to use a pocket calculator.

Testing Time Length

An ASE test session is four hours and fifteen minutes. You may attempt from one to a maximum of four tests in one session. It is recommended, however, that no more than a total of 225 questions be attempted at any test session. This will allow for just over one minute for each question.

Visitors are not permitted at any time. If you wish to leave the test room, for any reason, you must first ask permission. If you finish your test early and wish to leave, you are permitted to do so only during specified dismissal periods.

Monitor Your Progress

You should monitor your progress and set an arbitrary limit to how much time you will need for each question. This should be based on the number of questions you are attempting. It is suggested that you wear a watch because some facilities may not have a clock visible to all areas of the room.

Registration

Test centers are assigned on a first-come, first-served basis. To register for an ASE certification test, you should enroll at least six weeks before the scheduled test date. This should provide sufficient time to assure you a spot in the test center. It should also give you enough time for study in preparation for the test. Test sessions are offered by ASE twice each year, in May and November, at over six hundred sites across the United States. Some tests that relate to emission testing also are given in August in several states.

To register, contact Automotive Service Excellence/American College Testing at:

ASE/ACT
P.O. Box 4007
Iowa City, IA 52243

4 An Overview of the System

Damage Analysis and Estimating (Test B6)

The following section includes the task areas and task lists for this test and a written overview of the topics covered in the test.

The task list describes the actual work you should be able to do as a technician that you will be tested on by the ASE. This is your key to the test and you should review this section carefully. We have based our sample test and additional questions upon these tasks, and the overview section will also support your understanding of the task list. ASE advises that the questions on the test may not equal the number of tasks listed; the task lists tell you what ASE expects you to know how to do and be ready to be tested upon.

At the end of each question in the Sample Test and Additional Test Questions sections, a letter and number will be used as a reference back to this section for additional study. Note the following example: **D1.**

Task List

D. Vehicle Construction (6 questions)

Task D1 **Identify the type of vehicle construction (space frame, unibody, full frame)**

Example:

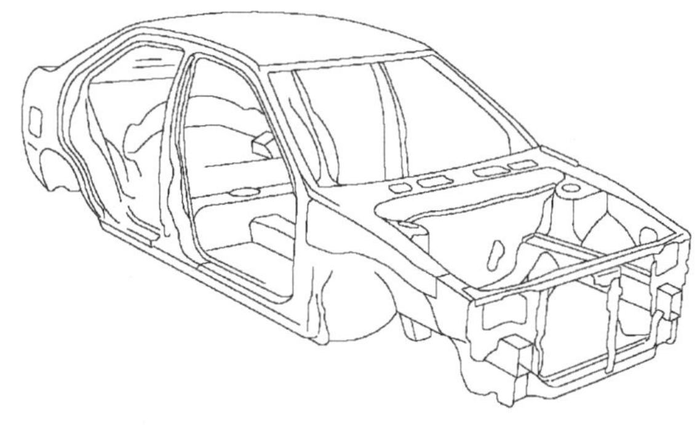

29. The figure above shows a:
    A. space frame.
    B. unibody.
    C. full frame.
    D. partial frame. (D1)

Question #29
Answer A is wrong.
Answer B is correct. The above figure shows a unibody vehicle construction.
Answer C is wrong.
Answer D is wrong.

Task List and Overview

A. Damage Analysis (10 Questions)

Task A1 Position the vehicle for inspection.

Prior to estimating the damage on a vehicle, the vehicle needs to be in a well-lighted area in which you can easily identify all damaged components. The area should be free of obstructions and have the necessary room required to adequately inspect the affected vehicle. It is also recommended that the inspection location be equipped with the capability to lift the vehicle to inspect for any underbody or suspension damage.

Task A2 Prepare vehicle for inspection by providing access to damaged areas.

In order to fully estimate the actual damage, viewing access must be gained by removing components that may inhibit the estimator's view of any damaged areas. This may require partial disassembly to fully inspect and estimate the damage. It is also advisable that the vehicle be cleaned of dirt, road salt, or road grime that may inhibit the estimator's ability to view vehicle damage.

Task A3 Analyze damage to determine appropriate methods for overall repairs.

When analyzing damage to determine appropriate methods for overall repairs, an estimator must evaluate the cost of repairing the affected area versus the cost of replacing it. Additionally, the estimator must evaluate the quality of the repair versus panel replacement to ensure the integrity, safety, and longevity of the repair.

Task A4 Determine the direction, point of impact, and extent of direct and indirect damage.

When estimating collision damage, the estimator needs to determine the direction and point of impact. The point of impact will be direct damage. This information can be gathered from a visual inspection, customer dialogue, or a police report. A visual inspection should include panel fit and finish. Check flexible surfaces for stress cracks. The interior should be examined for indirect damage. Indirect damage may be found anywhere on the vehicle, except the point of impact.

Task A5 Identify and record pre-existing damage.

Pre-existing damage, such as rust or door dings, may be repaired when repairing the current damage. This will improve the vehicle's condition from before the current accident. Betterment could be charged to the vehicle owner by the insurance company. Betterment charges apply to situations where the vehicle is improved from its condition prior to the loss while repairing the current damage; however, some situations may not apply (for example, minor paint damage near the current damage may be easier to repair by painting a whole panel instead of blending).

Task A6 Determine the cost-effectiveness of the repair and determine the approximate vehicle retail, salvage, and repair value.

The vehicle should be repaired in the most cost-effective manner that will not compromise integrity, safety, or longevity. The estimator should consult a used car value guide to determine the vehicle's retail and wholesale value. If the repair cost exceeds the retail value plus salvage value, the vehicle is considered a total loss and is usually not repaired. "Total loss" vehicles have salvage value, depending on the severity of damage.

Task A7 Perform visual inspection of structural components and members; determine if repair or replacement is required.

Because structural members support the vehicle and protect the vehicle's occupants, the diagnosis and proper repair of structural panels is of the utmost importance. A careful visual examination of the damaged high-strength steel structural parts is required. If a structural part is kinked, distorted more than 90 degrees over a short radius, the strength of the part is lost. If the structural part is bent and not kinked, distorted less than 90 degrees over a short radius, it may be able to be repaired and still have all of its strength. As a rule, kinked structural members must be replaced, but bent structural members may be carefully repaired.

Task A8 Identify structural damage using measuring tools and equipment.

The following are used to measure for structural damage:

1. Tram gauge—to measure length, width, and diagonal
2. Self-centering gauges—to measure centerline and height
3. Laser—to measure length, width, diagonal, centerline, and height. This type of system may also have a computer to analyze the damage information.

The following equipment may be used to measure damage, but its primary use is during vehicle repair, not estimating:

1. Dedicated bench—Fixtures indicate vehicle damage.
2. Universal bench—Damage indicated by pointers.

Vehicle dimension guides may be used to determine what the undamaged dimensions should be.

Task A9 Perform visual inspection of non-structural components and members; determine if repair or replacement is required.

Non-structural parts, such as fenders, doors, hoods, deck lids, quarter panels, and bumpers, should be carefully evaluated to determine if repair or replacement is required. If the repair cost exceeds the replacement cost, then the part should be replaced. Many times inexpensive parts, such as fenders, are replaced even if they are repairable, because repair cost exceeds replacement cost. More expensive parts, such as doors, may be repaired by replacing the door skin. Damaged bumper fascia may be repaired with adhesives. The estimator should check all doors, hoods, and deck lids for proper operation. Binding may indicate damage to hinges.

Task A10 Determine parts and components necessary for proper repair.

The estimator should use a collision estimating guide to determine what parts are needed for repair. The best practice is to start at the point of impact and work towards the center of the vehicle. The collision estimating guide can be used as a reference for part numbers and part location. An assembly made of several parts, such as a bumper, will have the individual parts listed in the guide.

Task A11 Identify type and condition of finish; determine if refinishing is required.

A vehicle's finish may be single stage (color only), basecoat/clearcoat or tri-coat (base color, intermediate pearl coat and clearcoat). To determine if a non-white color is a single stage or clearcoat, sand the damaged panel to see if the sanding dust is white or the color of the vehicle. If the dust is white, the vehicle has clearcoat. If the dust is the same color as the vehicle, the paint is single stage. The paint code, located on the vehicle, and the paint manufacturer's color book are used to determine if a white vehicle has clearcoat or if the vehicle has a tri-coat. The condition of the paint finish is best evaluated out of direct sunlight, inside a shop. Look for dull or peeling paint, overspray from a previous refinish or rust.

Most replacement panels will require refinishing. Fenders, hoods, doors, quarter panels, and all other OEM painted panels will be supplied as unpainted replacement parts. The only exception is that some plastic parts, such as bumper covers, may be molded in color.

Task A12 Identify suspension, electrical, and mechanical component damage.

While estimating the physical damage, the estimator should be aware of and record (further diagnosis may be required) any known or possible damage to suspension components, electrical components and wiring, and any other mechanical components that could be affected. The vehicle repair manual should be utilized to determine proper procedures.

To check for steering and suspension damage on a front-wheel drive vehicle, perform these quick checks:

1. Center steering wheel by turning the wheel in one direction to lock.
2. Turn the wheel in the other direction to lock. Count the number of turns.
3. Divide the number of turns, lock to lock, by 2. This number is half the lock-to-lock number of turns.
4. Turn the wheel half lock-to-lock number. Mark the top of the steering wheel with a piece of masking tape.
5. Check the steering wheel. If it is not centered, suspect steering box damage.
6. Check the front wheels. If they are not pointing straight ahead, suspect steering arm damage.
7. Jounce the front suspension by pushing down and releasing the front bumper. Watch the tape on the steering wheel.
8. If the wheel turns as the suspension is jounced, suspect the steering rack misalignment.
9. Set up a reference point and spin a wheel to check for a bent wheel or bent axle.
10. Measure rotor-to-strut distance to determine if a strut is bent.

Task A13 Identify safety systems damage and related service requirements.

Seat belts must be inspected. Consult the manufacturer's recommendations for replacement criteria. Overstressed, frayed, or damaged seat belts must be replaced.

Different manufacturers have different requirements for inspection and repair of air bag systems after a deployment. Some manufacturers, but not all, require all sensors be replaced. Deployed air bags cannot currently be rebuilt or reconditioned.

Task A14 Identify interior component damage.

A sudden stop must disperse all the kinetic energy of a moving vehicle, which includes the occupants and other objects inside the vehicle. The kinetic energy is determined by the vehicle's weight and speed. The inertia force of stopping the occupants suddenly may strain or damage the seat and seat belt parts. The damage may consist of frayed or stressed seat belts. They must be examined for damage. The inertia force of stopping loose objects inside the vehicle, both passenger compartment and trunk, may cause additional damage and is considered part of the loss.

Task A15 Identify damage to add-on accessories and modifications.

While estimating damage, the estimator must make note of any add-on accessories or vehicle modifications. These must all be accounted for on the estimate as they are non-standard and therefore are not included operations in published labor guides.

B. Estimating (13 Questions)

Task B1 Determine and record customer/client and insurance information.

Prior to making a complete physical damage appraisal, the estimator must record the customer information first:

1. Name
2. Home address

3. Home phone number
4. Work phone number

Insurance information should also be obtained. The following insurance information is useful:

1. Insurance company name
2. Claim representative's name
3. Claim representative's phone number
4. Claim number

Task B2 Identify vehicle identification number (VIN), make, model, year, production date, body style, trim level, paint code, engine, transmission, mileage, and license plate information.

The estimator must also complete vehicle information: vehicle identification number (VIN), make, model, year, production date, body style, trim level, paint code, engine, transmission, mileage, and license plate information. Identify all options and write them down. Some options may not appear to be involved with the repairs, but a complete list will help if they become important later when working with collision repair data. All options are also important when an actual cash value (ACV), the expected selling price of a vehicle prior to the loss, is needed.

Task B3 Identify options, conditions, accessories, modifications, and safety systems.

Identify options, conditions, accessories, modifications, and safety items. All of these items are necessary at times in determining parts costs, repair procedures, and additional labor that may be necessary to complete the repair. Sometimes it is necessary to have option codes to be positive that the correct safety items are installed.

At least three copies of the written estimate should be made: one is kept by the shop, one is given to the insurance company; and the other is given to the customer. An estimate is an approximate bid for a given period of time, usually for 30 days. The reason for a given time period is that part prices change and damaged parts can deteriorate.

Task B4 Apply appropriate estimating nomenclature (terminology).

A few common abbreviations and terms are:

- R&R—to remove and replace.
- R&I—to remove and install.
- Overhaul—to remove an assembly from the vehicle, and disassemble, clean, inspect, replace necessary parts, reassemble, install, and adjust (alignment excluded).
- Included operations—operations that can be performed individually, but are also part of another operation.

Task B5 Apply appropriate parts nomenclature (terminology).

Automotive repair terminology refers to parts on the right side and left side of the vehicle as viewed from the driver's position. Parts bolted into a unit are referred to as an assembly. Interchangeable parts are parts that are the same on two or more types of vehicles, such as hoods, fenders and doors. Parts are identified by a part number. Assemblies are made of several parts, such as a bumper. A bumper assembly consists of bumper fascia (cover), bumper reinforcement, and impact absorber.

Task B6 Determine and apply appropriate estimating sequence.

Each collision estimating guide for domestic vehicles covers one vehicle manufacturer, such as Ford. For imported vehicles there is one region of origin, such as Asian, per guide. Each guide contains several vehicle models, such as Escort or Camry. Each model is broken down into major assemblies, such as fender or engine. Commonly damaged parts are listed. The assemblies are arranged from the front of the car to the rear of the car. Because of the number of parts involved, the estimator should use an estimating sequence that follows the guide. For a front hit, begin at the rear and work forward. Work through each assembly section.

Task B7 Utilize estimating guide procedure pages.

The procedure (P) pages of a collision estimating guide list included and not included labor operations for each major vehicle assembly. For example in fender replacement, side marker lamp and turn signal lamp replacement are included, not included in fender replacement are front bumper removal, refinishing, decal replacement or hole drilling. The estimator must be familiar with the "P" pages because these pages list what operations are included in the labor allowances noted in the part assembly sections. If, for example, a fender is to be replaced on a vehicle and the front bumper must be removed to allow access, the estimator must add additional labor time to cover the bumper removal. Otherwise, the bumper removal will be done for free.

Task B8 Apply estimating guide footnotes and headnotes as needed.

The head notes of each major assembly list Refinish, R&I, and O/H times. The foot notes list included/not included items that are different from the "P" pages. The head notes and foot notes must always be heeded when writing an estimate.

Task B9 Estimate labor value for operations requiring judgment.

Slightly damaged panels may be repaired and not replaced. This repair time must be figured by the estimator. This is also known as judgment time. Judgment time must include time for the repair technician to analyze the damage, plan the repair, anchor the vehicle if needed, rough out with hydraulic equipment or hand tools, remove paint, metal finish, or apply filler, cure filler, sand filler, prime and block sand. If panels must be removed for access, the judgment time should reflect the additional labor.

Task B10 Select appropriate labor value for each operation (structural, non-structural, mechanical, and refinish).

The labor allowance for operations is dependent on the vehicle year, make, model, body style, and options. The estimator must be able to read through the collision estimating guide and select the correct labor allowance for each vehicle aspect.

Task B11 Select and price OEM parts, verify availability.

The collision estimating guides list the part numbers and prices for OEM parts. The part numbers for left and right parts, such as fenders, may be listed together. In that case, if the number listed is 123456-7, the right-hand part is listed first, 123456, and the left-hand part is listed second, 123457. In some cases, parts may be interchangeable. In other cases, parts may no longer be available from OEM. They would be indicated as discontinued. Occasionally, parts do not have a price listed. In that case, the estimator must call the OEM dealer for the part price.

Task B12 Select and price aftermarket parts; verify availability.

Non-OEM parts are also known as aftermarket parts. Aftermarket parts are usually the commonly damaged parts such as fenders, bumpers, hoods, and lights. There is an ever increasing number of aftermarket parts available. Prices and part numbers are listed in catalogs supplied by the jobbers or mail order warehouses. Even though a part number may be listed, the part may not be available. The estimator should call the supplier to verify if the part is actually available. Aftermarket parts usually cost less than OEM parts.

Task B13 Select and price salvage/used parts; verify availability and condition.

Salvage or used parts are usually available as assemblies, such as a used door includes the door shell, regulator, glass, and weather strips. Used parts are priced at approximately half of the OEM price. Availability of used parts may be limited. It stands to reason that more common vehicles will have greater used part availability than less common vehicles. The condition of used parts can vary. Common problems with used parts include previous repairs, rust, and dents. The price, availability, and condition of used parts can be verified by calling the salvage yard. Usually the estimator adds a 20–25% markup to the price of the used part to determine the price the customer pays. If the used part is in poor condition, the estimator should negotiate with the salvage yard to determine a price that covers the proper repair of the part.

Task B14 Select and price remanufactured, rebuilt, and reconditioned parts; verify availability.

Some parts are available as reconditioned or rebuilt. Commonly reconditioned parts are bumpers, both plastic and metal. Reconditioned metal bumpers are sometimes called rechromed. Rebuilt parts are usually engine parts such as water pumps and alternators or motors such as a power window regulator. The prices are listed in catalogs provided by jobbers or mail-order warehouses. The availability of parts should be verified by calling the supplier.

Task B15 Determine price and source of necessary sublet operations.

Sublet operations are repairs that a body shop is unable to do in-house and must have another shop perform. Some common sublet operations include radiator repair, A/C recharge, towing, front end alignment, and air bag replacement. The shop usually has an established supplier for sublet operations. The sublet shop may charge the body shop less than the standard fee for services. The body shop then charges the customer the usual fee. If this sublet shop charges the body shop the usual rate, then the body shop marks up the sublet charge by 20–25%.

Task B16 Determine labor value, prices, charges, allowances, or fees for non-included operations and miscellaneous items.

Aftermarket parts may require more time to fit than OEM parts. The estimator should take into account this extra labor time. Aftermarket accessories, such as running boards or stripes, will have no labor time or price listed in the collision estimating guide. The estimator must determine the price of these parts and determine labor allowance to install.

Task B17 Recognize and apply overlap deductions, included operations, and additions.

When two panels are replaced and they have a common seam, the replacement of one will make the replacement of the other easier. This is called overlap. Overlap is deducted from the labor time of one panel. For example, if a quarter panel and rear body panel are replaced, the labor time to replace a quarter panel on a '93 Mustang is 16.0 hours. The labor time to replace a rear body panel is 7.5 hours. The collision estimating guide says to deduct 1.5 hours from the rear body panel labor for each quarter panel removed. In this case, the labor time for the quarter panel is 16.0 hours and the rear body panel is 6.0 hours. Overlap also applies to refinishing.

Included operations are labor times that are part of an operation. These are listed in each collision estimating guide section. The "P" pages list included operations. More specific included operations are listed in collision estimating guide sections. For example in the "P" pages, front bumper R&I is not included in radiator support replacement. However, on a '98 Escort in the front inner structure section, R&I front bumper assembly is listed as included in radiator support R&R. By checking the "P" pages and notes in each section, the estimator can determine what labor is included.

Additions must be made if an operation is required but not listed as included. For example, pickup truck cab corner replacement does not include bed R&I. If the bed needs to be removed for access when replacing a cab corner, an addition for bed R&I

must be made. Refinish labor time is listed in collision estimating guides. There are additions and overlap. For example, if a fender is to be replaced, the guide may list 2.5 hours to repaint the outside of the fender. The fender edges need to be painted also, so the guide lists 0.5 hours to edge the fender. Total refinish time would be 3.0 hours. When more than one panel, not including painted bumpers, is refinished, an overlap deduction should be made for each panel. If the painted panels are adjacent, a 0.4 hour deduction is made. If the panels are not adjacent, a 0.2 hour deduction is made. For example, if the right fender, right door and right quarter panel are to be repainted, the refinish time for the fender is 2.5 hours, the door is 3.0 hours and the quarter panel is 3.0 hours. The total time is 8.5 hours. An overlap of 0.8 is taken, 0.4 hours for the door and 0.4 for the quarter panel. So the refinish time is calculated at 7.7 hours.

clearcoat time is added to the refinish time at the rate of 0.4 per refinish hour for the first panel, plus 0.2 per refinish hour for additional panels. In the side repair example, the clearcoat time would be 2.5 x 0.4 = 1.0 for the fender, the first major panel plus 3.0 x 0.2 = 0.6 for the door and 3.0 x 0.2 = 0.6 for the quarter panel. The total clearcoat time would be 1.0 hour + 0.6 hour + 0.6 hour or 2.2 hours. Three-stage paint is figured in a similar manner except the numbers are 0.7 per refinish hour for the first panel plus 0.4 per refinish hour for additional panels. Two-tone is figured at 0.5 per refinish hour for the first major panel plus 0.3 per refinish hour for additional panels. Blended panels are figured at 0.5 per refinish hour.

Task B18 Determine additional material and charges.

The estimator must make allowances on the estimate for any additional materials or charges that may be necessary in order to complete the repairs. This would include, but is not limited to, body repair materials, rust-proofing, solvents, fasteners, and cleaning materials. It would not include items such as emblems or decals.

Task B19 Determine refinishing material and charges.

Refinish material costs can be figured two ways, each uses the number of refinish hours. In one method, the number of refinish hours is multiplied by a dollar amount. For example, if the vehicle requires 6 hours of refinish time and the shop charges $20.00 per refinish hour, the refinish material amount is 6 hours x $20/hour or $120.00. There are reference books available that use the paint code and refinish hours to determine the refinish material costs. The estimator simply looks up the paint code and the appropriate refinish hours. The refinish material costs are listed.

Task B20 Determine sectioning procedures where appropriate and establish labor values.

When replacing major panels, the estimator must refer to original equipment manufacturer's (OEM) repair manuals and other reputable sources to determine if and at what location a panel may be sectioned and if that procedure changes the labor value of the base operation. It is imperative that the procedure is approved by the OEM so that the structural integrity of the vehicle is not compromised. Collision estimating guides often list sectioning locations and labor allowances. The OEM manuals may list structural sectioning locations, but no labor time will be given. Structural members that are bent and are not kinked, may be repaired and retain full design strength. Frame straightening consists of anchoring the vehicle to immobilize it, then pulling and stress relieving the damaged members back into the proper location. The procedure would be to anchor the vehicle, measure, plan repair sequence, attach clamps and/or chains, hook up pulling equipment, apply tension, spring hammer to relieve stress and remeasure. Once frame alignment is achieved, the clamps, chains, and anchoring are removed. The estimator must allow for the labor time that is necessary to perform all of these tasks.

Task B21 Determine structural measurement requirements; diagnose, and establish labor values.

Frames can be measured with gauges, benches, lasers, and computers. Vehicles with structural damage will need to be diagnosed to determine the location and extent of the structural damage. The estimator must make a labor allowance that indicates the time to measure and diagnose a damaged frame. No set criteria has been established, due to the wide variety of measuring systems that are available. The estimator must make measuring labor allowances based on his experience with the shop's measuring equipment.

Task B22 Determine necessary structural straightening equipment, set up procedures, and establish labor values.

Frame repair equipment can be a floor pot system with towers, a bench on which the vehicle is mounted, or a rack on which the vehicle is placed. All systems include devices to anchor the vehicle. Pulling is done with hydraulic towers or arms. Set up times vary greatly. A shop may have a combination of systems, using a simple floor pot system for light damage and a bench for heavy damage. The estimator must establish guidelines to determine how much set up time is required to position the vehicle, anchor the vehicle and position the towers.

Task B23 Apply math skills to establish charges and totals.

The estimator must be able to add, subtract and multiply. Estimate totals for labor, parts, materials and taxes are a combination of all of these math skills.

Task B24 Interpret computer-assisted and manually written estimates; verify the system has current information.

Estimate forms for manually written estimates often have columns marked for repair, replace, and sublet. As the estimator writes the estimate, he checks the appropriate column. Abbreviations are often used to indicate body panels, such as quarter panel is abbreviated as q.p. Computer-generated estimates also utilize abbreviations. In some computer generated estimates, the type of part—new, OEM, new aftermarket, or used—is indicated. The words "repair" and "replace" are printed before the panel name, such as repair left fender.

Because repair prices often change, current information is a necessity. For manually written estimates, the date of the collision estimating guide used to write the estimate should be checked. If the date is not current, the part price information is suspect. The same is true for computer generated estimates. The date of the data should be checked.

Task B25 Identify procedural differences between major computer-assisted and manually written estimating systems.

There are three major information companies that provide computer generated estimate information. The information provided—part prices, part numbers, and labor allowances—is similar. However, the way the data is accessed, entered, and printed is different. When writing a computer-generated estimate the procedure varies depending on the information company.

For manually written estimates there are two major companies providing collision estimating guides. These manuals contain similar information, part prices, part numbers, and labor allowances. However, the information is arranged differently.

Task B26 Identify procedures to restore corrosion protection; establish labor values.

When utilizing OEM repair manual recommendations, the estimator must determine if corrosion protection is needed and make allowances for corrosion protection in the affected area in order to return the vehicle to pre-accident condition.

Corrosion protection should be applied when a part such as a fender or door is replaced. If a replacement panel is welded, such as a quarter panel, corrosion protection is very important. A weld is a corrosion hot spot. Weld prep should consist of cleaning the weld, applying epoxy primer and seam sealer. Then spray a wax-based corrosion protection. The same is true on sectioned structural panels. Labor time on non-welded replacement panels should include time to set up the spray equipment, mix the materials, apply the materials, and clean up the equipment.

Task B27 Determine appropriate application of betterment/depreciation to parts and allowances as necessary.

In order to satisfy insurance requirements, the estimator must deduct betterment/depreciation. Generally, tires are measured for remaining tread and depreciated based on tire used. Wearable suspension parts are generally depreciated based on mileage up to 50 percent. Insurance adjustments can apply betterment/depreciation to almost any component on a vehicle.

The insurance company is obligated to return the customer's vehicle to pre-accident condition, no better and no worse. Betterment is subtracted if the vehicle has worn parts or existing body defects, such as faded paint. By replacing panels, the vehicle would be in better shape than its pre-accident condition. The insurance company does not want to make the owner's vehicle better than it was before the accident.

C. Legal and Environmental Practices (4 Questions)

Task C1 Recognize regulatory obligations.

Collision repair facilities are subject to federal, state, and local regulations. The regulatory agencies include the Environmental Protection Agency (EPA) and the Occupational Safety and Health Administration (OSHA). The EPA regulates the air, land, and water pollution sources. For collision repair facilities, this includes discharge of volatile organic compounds (VOCs) and hazardous waste disposal. OSHA regulates the workers' on-the-job safety. Each shop is required to inform the workers about the hazards of the chemicals that they will work with. These chemicals include solvents, paint, and fillers.

Task C2 Recognize contractual and warranty obligations.

An estimate lists the parts that will be repaired or replaced. The collision repair facility must perform all operations as specified on the estimate unless the customer agrees to additions, deletions or substitutions after the work begins. To do less than the specified amount of repairs without the customer's permission, is not correct. An example would be if the shop repairs a quarter panel instead of replacing it and the shop still charges the customer for the new part.

Collision repair facilities may offer a warranty on the repair. The length of the warranty is determined by the shop. The customer should receive the warranty in writing.

Task C3 Recognize the legal obligations to restore the vehicle to pre-loss condition based on established industry standards and vehicle manufacturer's recommendations.

When a collision repair facility is hired to restore a damaged vehicle to pre-loss condition, the shop must insure that all of the repairs do, indeed, restore its pre-accident condition. For example, if a vehicle frame is damaged and the shop repairs the frame damage, then the frame must be restored to its original dimensions. If the repaired vehicle is involved in another collision, the frame must react in the same manner as a never-damaged frame. This pre-loss condition also includes panel alignment, safety component function, part operation, paint durability and color match. However, the shop is not obligated to repair a part that was non-functional before the loss and is not included on the estimate. For example, if a left door window regulator did not work before the vehicle was damaged on the right side and the repair of the left window regulator was not listed on the estimate, then the shop is not responsible if the window does not work. If additional damage, not listed on the estimate, was found while repairing the vehicle, the owner or the insurance company should be contacted so that they may inspect the damage and authorize additional repairs.

D. Vehicle Construction (6 Questions)

Task D1 Identify type of vehicle construction (space frame, unibody, full frame).

Vehicle construction can be unibody, full frame or space frame. A unibody has no detachable frame. The vehicle is constructed of welded together panels. Most sub-compact, compact, and intermediate sized vehicles are unibody. A full frame vehicle is constructed by bolting a body to a separate frame. Most pickup trucks, sport utility vehicles (SUVs) and some large passenger vehicles have a full frame. A space frame is similar to a unibody in that it is made of welded panels. However, in a space frame, outer body panels do not contribute to the strength of the unibody. Lumina type vans and Saturn models are examples of space frames.

Task D2 Recognize the different damage characteristics of unibody and frame-type vehicles.

The unibody is designed to absorb the impact or crush during a crash. The energy of the collision is absorbed by convolutions in the structural members. The collision energy is gradually dissipated as it is transferred along the structural parts. In a full frame vehicle, the collision energy is not absorbed in the same manner as a unibody; the energy is transferred. A full frame is stronger than a unibody, however. A full frame vehicle can have the following types of frame damage: diamond, twist, sag, mash and side sway. A unibody vehicle may have these types of damage: twist, sag, mash and side sway.

Task D3 Determine repairability of impact energy absorbing components.

Some vehicles are equipped with energy-absorbing bumper shocks. These shocks absorb the force of a minor (5 mph) impact. After the minor impact, the absorber returns to its normal shape. Greater than a 5 mph impact may damage the bumper shock.

Repairable damage consists of a minor bend on the mounting plate. Non-repairable damage consists of a bent tube, a fluid-leaking tube or a broken tube.

Other vehicles are equipped with a thick styrofoam pad in between the bumper cover and the bumper reinforcement. The bumper cover should be removed if damage is suspected. The only way to see if the styrofoam impact absorber is damaged is to remove the bumper cover. A broken or dented styrofoam impact absorber should be replaced.

Task D4 Identify steel components and repair/replacement procedures.

A vehicle may be made of as many as three types of steel. Most of the sheet metal panels on a vehicle are made of mild steel. Mild steel may be welded, heated and straightened. High-strength steel is used on some or all structural members. High-strength steel must be repaired with special techniques. Bends in high-strength steel may be carefully repaired. A kink in high-strength steel calls for panel replacement or sectioning. A special technique for repair of high-strength steel include not heating the steel above the manufacturer's recommended temperature. Some outer panels, such as hoods, deck lids, and fenders may be made of high-strength steel as well as structural parts.

Ultra high-strength steel is used on door crash beams. This steel can not be repaired, if damaged. The component must be replaced.

Task D5 Identify aluminum components and repair/replacement procedures.

Some bumper reinforcements are made of aluminum. Minor bends may be straightened. Tears may be welded with a specially-equipped MIG welder or a TIG welder. Some body panels, such as hoods or deck lids are made of aluminum on certain models. Dents on these panels may be repaired with metal finishing and body fillers. At some time in the future, the entire unibody may be made of aluminum.

Task D6 Identify plastic components and repair/replacement procedures.

Plastic panels may be rigid or flexible. Rigid panels may be bumpers, fenders, doors, van side panels, hoods, roofs, and rear hatches. Rigid plastic may be SMC, RRIM, PC or ABS. Minor damage on rigid panels may be repaired with adhesives. Outer panels may be replaced on rigid plastic doors. Plastic van sides may be spliced. Flexible plastic include bumper covers, and ground effects. Only certain types of plastic can be welded. All types of plastic may be repaired with adhesives, however, plastics containing olefin must have a special adhesion promoter applied. Stretched or misshaped flexible plastic may be reshaped with heat.

Task D7 Identify vehicle glass components and repair procedures.

Remove and replace (R&R) of a windshield indicates a new windshield will be installed. Remove and install (R&I) of a windshield indicates the same windshield will be re-installed. This will make a difference in removal. More care and attention must be taken to remove an unbroken windshield for reuse than a broken windshield for disposal. More labor is needed for the R&I operation.

Stationary urethane-set glass has a minimum drive-away time and a full cure time. Prior to the minimum drive-away time, the urethane has not cured to a strength safe enough to perform as designed in another collision. During this curing time, a sudden force, such as interior pressure from slamming a door, may dislodge the newly set glass. It is often economical to have an automotive glass specialist do the glass work because they come to the collision shop to do installations.

Quarter panel or encapsulated glass have a molding that is set with the glass. Removal without damage is difficult. Always note in the estimate that there is a chance of breakage. Moveable glass may be bonded or bolted into the window regulator.

Task D8 Identify add-on accessories and modifications.

Add-on accessories include ground effects, fog lights, running boards, fender flares, bed liners, and graphics. These accessories are not included in collision estimating guides. When these parts are damaged, the estimator must consult aftermarket suppliers for replacement part availability.

E. Vehicle Systems Knowledge (10 Questions)

1. Fuel, Intake, Ignition, and Exhaust Systems (1 Question)

Task E1.1 Identify major components.

Estimators must be able to identify the major components of the fuel, intake, ignition, and exhaust systems so that the proper components will be ordered; for example, on newer vehicles, the vapor canister may be located near the fuel tank, and many power train control modules (PCM) are now located in the engine compartment.

Task E1.2 Identify component function.

Estimators must understand the functionality of the major components of the fuel, intake, ignition, and exhaust systems so that they can make a proper decision on whether they might need further diagnosis after the collision repairs (for example, no start). Some vehicles have an inertia switch that shuts off fuel after an impact, and most vehicles today have an in-tank electric fuel pump. Many vehicles have a sensor behind the harmonic balancer that creates a signal for the spark and fuel. A knock sensor is designed to retard spark after the engine begins to knock and therefore should not cause a no start condition.

Task E1.3 Identify OEM component service requirements.

An estimator needs to know the OEM serviceability of the various fuel, intake, ignition, and exhaust systems components so that the estimate will be as accurate as possible, and also needs to be prepared for special procedures that may be necessary after the installation of some components. Fuel additives should not be added unless recommended by the manufacturer. Vacuum leaks on many newer vehicles cause a high idle while on older vehicles they can cause a rough idle.

2. Suspension, Steering, and Powertrain (2 Questions)

Task E2.1 Identify components.

Steering systems can be rack-and-pinion or parallelogram. The rack-and-pinion system consists of a steering rack and tie rods. A parallelogram steering system consists of a steering gear box, Pitman arm, center link or drag link in some applications, tie rods, idler arm and tie-rod ends. Front suspensions can be McPherson strut consisting of a strut, spring and lower control arm, or short arm, long arm consisting of upper and lower control arms, spring, and shock absorber.

Rear suspension for front wheel drive vehicles include struts and springs. Rear wheel drive rear suspension may consist of leaf spring or coil springs. Both front and rear suspensions may include stabilizer bars. The power train consists of an engine, transaxle, and half shafts on a front wheel drive vehicle. A rear wheel drive train consists of an engine, transmission, transfer case (on four-wheel drives), drive shafts, and rear axle (front axle also on four-wheel drives).

Task E2.2 Identify component function.

The steering system transfers the turning of the steering vehicle into turning of the front wheels. These components are subject to wear as well as damage. To function properly, all components must be aligned and any "play" must be within the tolerances.

The suspension system's function is to keep the wheels on the road and absorb road shock. These components are subject to wear and damage. To function properly, the parts can not be damaged and all "play" must be within tolerances. The power train function is to generate power and transfer it to the drive wheels. To function properly, the drivetrain must be in alignment with the frame. No components should be damaged.

Task E2.3 Identify OEM component service requirements.

Thrust angle is the direction of the rear wheels as compared to the vehicle centerline. If the rear wheels are not tracking straight with the vehicle centerline, the front wheel alignment settings must be changed from specifications to compensate; otherwise, they will compromise both performance and tire wear. The thrust angle must be correct first before the front wheel angles can be properly set. Camber is the inward or outward tilt of the wheel as viewed from the front. Caster is the forward or rearward tilt of the steering axis as viewed from the side. Toe is the inward or outward pointing of the wheels from the center line.

In all cases, bent or otherwise damaged steering and suspension components are replaced and not straightened. In some cases, such as a rear axle housing, power train components may be straightened, if the damage is minor.

Typically an alignment is only necessary when replacing an adjustable steering component such as a tie rod, not an idler arm or center link.

3. Brakes (1 Question)

Task E3.1 Identify components.

Disk brakes found on the front or rear, consist of a rotor, caliper and pads. Drum brakes, found on the rear only, consist of a drum, wheel cylinder, and shoes. Brake operation is controlled by the master cylinder. Antilock brakes (ABS) may be one channel—rear wheels controlled together, two channel—rear wheels controlled separately, three channel—rear wheels controlled together and each front wheel controlled separately or four channel—all wheels controlled separately. An ABS system consists of a tone ring, speed sensor, and modulator.

Task E3.2 Identify component function.

When the driver steps on the brake pedal, hydraulic fluid from the master cylinder causes the brakes, either pads or shoes, to press against either the rotor or the drum. For the system to function, the brake lines must be able to hold the pressure. A loss of pressure in the system is indicated by a glowing red brake warning light. An ABS system is capable of self-checks and will indicate a system malfunction by a glowing ABS warning light. The brake system, if undamaged, will operate if the ABS warning light glows. However, the antilock brake function will be lost.

Task E3.3 Identify OEM component service requirements.

The technician should check OEM requirements for brake component service. In most cases, undamaged components, with wear that is within tolerances, such as brake shoes, pads, disks or drums may be reused if the brakes must be disassembled for repair. If a brake line is disconnected, as in the replacement of a caliper or brake line, the brake system must be bled. An allowance must be given on the estimate to cover brake bleeding.

4. Heating, Engine Cooling, and Air Conditioning (2 Questions)

Task E4.1 Identify components.

The heating system consists of heater hoses, heater core, and controls. The air condition system consists of a condenser, evaporator, accumulator or drier, compressor, and expansion valve.

Task E4.2 Identify component function.

The heating system uses the heat from the hot coolant to blow the warm air into the vehicle. Because the system is under pressure, all lines must be intact. The air conditioning functions because as the refrigerant changes its state from a liquid to a gas, it absorbs heat. The A/C system is a continuously circulating, pressurized system in which the refrigerant is changing from a liquid to a gas in the evaporator (absorbing heat), then changing from a gas to a liquid in the condenser (releasing heat). The refrigerant can be R-12 or R-134a. These two refrigerants cannot be interchanged. The type of refrigerant is specified on the vehicle. The compressor builds up system pressure. The drier or accumulator removes the harmful moisture from the system.

Task E4.3 Identify OEM component service requirements.

A damaged condenser should be replaced. The O-ring connectors should be replaced whenever the A/C lines are disconnected. All kinked or broken A/C lines should be replaced. If the A/C system has been open to the atmosphere for more than an hour, the drier should be replaced as well. Depending on the type of system, the proper refrigerant, R-12 or R-134a, must be used for recharge. Oil should also be added to the system as needed. If the system is under pressure before repairs, the refrigerant must be recovered and stored, to be reinstalled after repairs are made.

5. Electrical/Electronic Systems (1 Question)

Task E5.1 Identify components.

Electrical system components most commonly damaged in a crash include batteries, headlights, horns, taillights, turn signals, park lamps, power window regulators, alternators, power door lock actuators, electronic control modules and the wiring that connects them.

Task E5.2 Identify component function.

The battery stores the electrical energy. The wiring conducts the electrical energy to the various components. The lights and lamps use the electrical energy to produce light. The electrical motors, like the power window regulators, use the electrical energy to move the windows up and down. The alternator converts mechanical energy into electrical energy. The ECMs receive the information from various sensors and control the engine function.

Task E5.3 Identify OEM component service requirements.

Faulty electric or electronic components are usually replaced by collision repair facilities, and not rebuilt. For example, if an alternator is damaged in a crash, the alternator is replaced, rather than rebuilt by the shop. Broken wires may be repaired by crimping or soldering the wires together. A wiring diagram is helpful when tracking down an electrical problem. Follow OEM recommendations when using electrical power to test the circuits. Be sure to track down all possible sources of the problem before replacing parts. For example, if a power window does not operate in a damaged door, the wiring or regulator or both may be damaged.

6. Safety Systems (2 Questions)

Task E6.1 Identify components.

Vehicle safety systems include manual seat belts, automatic seat belts, and air bags. Manual seat belts consist of lap and shoulder belts or retractors and buckles. The user must buckle the seat belt.

Automatic seat belts operate by placing a shoulder belt over the seat occupant by way of a motor and rail system. The occupant must place the lap belt manually.

An air bag system consists of crash sensors, air bag module—driver's side or passenger's side—clock spring, lamp, and wiring.

Task E6.2 Identify component function.

Seat belts, both manual and automatic, function by restraining the occupant during a crash. To operate, the seat belt must be buckled and solidly attached at the anchoring point.

An air bag crash sensor detects the rapid deceleration of a frontal collision. The crash sensors send a signal to the air bag module. The air bag or air bags immediately inflate to absorb the occupant's impact. Then it deflates. If a vehicle is equipped with driver and passenger side air bags, both bags inflate, if activated, whether the passenger's seat is occupied or not. The crash sensors are designed to signal the air bag to inflate if the vehicle crashes at a speed of over 10 to 30 m.p.h., depending on the manufacturer. The air bag system will go through a system of self-checks each time the key is turned on.

When the self-checks are completed and the system is functional, the light goes off. A glowing air bag light indicates a system problem.

Task E6.3 Identify OEM component service requirements.

Seat belts should be checked after a collision. Some manufacturers recommend replacing all seat belts in use during a collision. The belts should be checked for cut or bowed webbing or broken threads. Manufacturers have specific recommendations on what parts of an air bag system need to be checked or replaced after deployment. In all cases, the air bag module must be replaced. It can not be reused. Undamaged crash sensors may be reused, if allowed by the manufacturer. In some cases, the clock spring and dash panel may need replacement.

7. Fasteners and Materials (1 Question)

Task E7.1 Identify fastener type.

There are many fasteners of different types used on vehicles today. There are specialized fasteners for virtually every connection, and you must always use the same fastener that the manufacturer originally installed. In addition to standard nuts and bolts, there are metric, flange nuts, castle nuts, wing nuts, jam nuts, flat washers, lock washers, fender washers, finishing washers, different kinds of screws, plastic clips, pins, and even adhesives.

Some fasteners, such as suspension bolts, are torque-to-yield. Once these bolts are installed and torqued, they can not be reused. Always replace any fasteners that are specified as torque-to-yield by the manufacturer.

Task E7.2 Identify fastener grade.

Always use the same grade bolt and nut as the manufacturer. For standard bolts, the more lincs (or dots) in the top of the head the stronger they are: for metric, the higher the number, the stronger the bolts are.

Task E7.3 Identify body repair and refinishing materials and supplies.

Body repair materials include body filler, plastic repair material, and MIG welding wire.

Body repair supplies include grinder disks and sandpaper. Refinishing materials include primer, surfacer, sealer, basecoat, clearcoat, hardener, and reducer. Refinishing supplies include sandpaper and rubbing compound. Materials remain on the vehicle once repairs are completed. Supplies do not.

Task E7.4 Recognize proper application and use of chemicals.

Refinishing materials have specific methods of application. Obtain a refinish manual from the paint supplier. Follow the manufacturers application instructions. Refinish materials are usually hazardous. Obtain, read, and follow all recommendations about what safety equipment must be worn when mixing and applying refinish materials.

F. Parts Identification and Source Determination (4 Questions)

1. New Original Equipment Manufacturer (OEM) (1 Question)

Task F1.1 Identify components.

Some vehicle components, such as air bag components, seat belt and other safety-related items are only available as OEM. Other components, such as sheet metal and drivetrain parts are available from the OEM as well as other sources.

Task F1.2 Identify component function.

The estimator must be able to recognize the function of various automobile components. For example, structural parts support the weight of the vehicle and absorb impact to protect the passengers. Safety components also protect the passengers.

Task F1.3 Justify repair or replace decision.

Damaged panels may be repaired if the cost of repair is less than the cost of replacement. Some panels are not repairable. A panel replacement is justified if the replacement cost is less than the repair cost. Some components, such as steering and suspension, must be replaced if damaged. The estimator must determine what is the most cost-effective way to return the vehicle to pre-accident condition. The vehicle safety, durability, fit and finish must not be compromised based on repair/replace OEM decisions.

Task F1.4 Determine availability.

Some OEM parts are no longer available. These are referred to as discontinued parts. Some OEM parts may not be readily available; these are called back-ordered. Parts may be back-ordered for a number of reasons. The back-ordered period may be a matter of days or as long as months. Other parts are in a regional warehouse and are available one or two days after ordering from the dealer. Commonly used parts are kept in stock at the dealer and are available immediately. In all cases, the estimator can call the OEM dealer to determine part availability.

2. New Aftermarket (1 Question)

Task F2.1 Identify components.

Many parts are available as aftermarket. These parts include sheet metal, glass, bumpers, drivetrain, steering and suspension components. The estimator should be able to recognize the various aftermarket components. For example, the packaging will not be OEM.

Task F2.2 Identify component function.

The estimator must be able to identify the function of the various aftermarket components. For example, an aftermarket fender's partial function may be to mount side marker lights. The estimator should be aware that an aftermarket fender may not have all of the necessary holes pre-drilled. In order to restore the damaged vehicle to pre-accident condition, the aftermarket parts must function the same as OEM parts.

Task F2.3 Justify repair or replace decision.

Repair and replacement judgments are based on the cost of damaged panel repair compared to panel replacement cost. Aftermarket parts cost less than OEM parts. The estimator should decide if aftermarket parts will restore the vehicle to pre-accident condition. If the use of aftermarket parts will not return the damaged vehicle to pre-accident condition, they should not be used.

Task F2.4 Determine availability.

Although many frequently replaced parts are available aftermarket, the less frequently replaced parts, such as frame rails, rocker panels, and center pillars, are not usually available as aftermarket. Some computerized estimating systems have aftermarket part databases. When a part is to be replaced, if it is available aftermarket, the part is listed on the estimate. The availability of the part should be confirmed by contacting the appropriate jobber. In some cases, aftermarket parts are listed in catalogs, but are not available.

3. Salvage (Used) (1 Question)

Task F3.1 Identify components.

Salvage or used, or like and kind (LKQ) are usually available as assemblies. For example, a used fender assembly may include the fender liner, moldings, and marker lights.

A good used assembly cost about half of what a new OEM part would cost. Some other terminology used with salvage parts:

- front clip—the entire front section of a vehicle, from the floor under the driver's seat forward
- rear clip—the entire rear section of the vehicle, from the floor under the rear seat rearward
- rear clip top—the rear clip plus the roof
- knee—the strut, knuckle, and control arm
- Salvage vehicles can be cut to give the customer any parts that are needed.

Task F3.2 Identify component function.

A used assembly, such a rear clip, can save considerable parts' cost. Some shops only use the needed parts from the clip and still the cost is less than OEM. For example, if a rear clip is obtained to repair a rear hit vehicle, the needed parts can be taken off of the clip. In this example, the rear bumper, rear body, panel, tail lights, deck lid, and a section of frame rail were used. The same is true for a front clip. The estimator should be aware that used parts may have repaired body damage or may even be replacement aftermarket parts.

Task F3.3 Justify repair or replace decision.

Used parts may be specified for older vehicles, vehicles that would be a total loss if new OEM parts were specified. In some cases, such as rear damage, there are fewer welds (windshield pillars, rocker panels and floor) and less labor in replacing a rear clip top than replacing each individual part OEM. The decision to replace with used parts should be based on the value and condition of the damaged vehicle. In some cases, used parts are specified for newer vehicles, possibly because the owner prefers OEM used parts to new aftermarket parts.

Task F3.4 Determine availability.

Used parts are available from salvage yards. Many salvage yards have computerized inventories. If a salvage yard does not have a part, they can often locate it at another salvage yard. The estimator can call the salvage yard to determine if a part is available. Some difficulty may be experienced if commonly damaged parts, such as front end parts, are needed for a late model vehicle.

4. Remanufactured/Rebuilt/Reconditioned (1 Question)

Task F4.1 Identify components.

Remanufactured, rebuilt or reconditioned parts commonly include bumper covers, bumpers, wheels, and electrical parts such as alternators. Remanufactured parts are repaired OEM parts.

Task F4.2 Identify component function.

The estimator should be able to determine the function of remanufactured parts. The use of remanufactured parts should not compromise the quality of the repair.

Task F4.3 Justify repair or replace decision.

The use of remanufactured parts will save money compared to OEM. The advantage of remanufactured parts is that the parts are damaged and repaired OEM. The fit should be the same as OEM. Remanufactured parts, like alternators, have a warranty. Remanufactured parts are usually specified for older vehicles. For example, if an older pickup truck has a damaged chrome bumper, a remanufactured or rechromed bumper may be installed. The rechromed bumper will save money compared to a new OEM. As always, the cost of panel replacement should be less than the cost of panel repair. The use of reconditioned parts may prevent a vehicle from being declared a total loss.

Task F4.4 Determine availability.

Not all parts are available as rebuilt. The availability can be checked by consulting a catalog. The supplier should also be contacted to verify part availability.

G. Customer Relations and Sales Skills (3 Questions)

Task G1 Acknowledge and greet customer/client.

When customers bring their vehicles to you, it shows that they already have trust in your abilities to repair their vehicles. A warm greeting will show the customer that you appreciate their trust and that you truly understand their situation, it will make them feel good that they have chosen a professional such as yourself.

Task G2 Listen to customer/client; collect information and identify customer's/client's concerns and needs.

It is imperative that you listen closely to all of the information that the customer gives you regarding the accident. This will accomplish two things: (1) it will develop a trust between you and the customer and (2) it will assist you in determining pre-existing and indirect damage that may have occurred.

Task G3 Establish cooperative attitude with customer/client.

By letting the customer know that they have your full attention and feel that you are there for them, they will feel comfortable working with you. Once the customer knows that you are there to answer any questions they may have and are willing to listen to their concerns and offer explanations to them, they will feel as though they are dealing with a trustworthy establishment.

Task G4 Identify yourself to telephone customer/client; offer assistance.

By identifying yourself on the initial phone contact with a prospective customer you start a trusting relationship, build confidence in your business, and establish the belief that they are working with a true professional. When a customer leaves their vehicle with you, it is appropriate to let the customer know that you are available to them at all times to discuss the progress of the repairs, explain the types of repairs and replacements to be done, and give them an updated completion estimate. This helps build a relationship between you and the customer. If the customer knows that they can call you at any time to check progress or ask a question, they will know that they have taken their vehicle to a quality establishment and trust that everything will be repaired properly.

Task G5 Use salutation skills (post greeting).

When a customer leaves with their vehicle it is always appropriate for you to thank them for their business and their trust in you to return their vehicle to pre-accident condition. This lets the customer know that you really do care about both them and their vehicle. Other methods of contact after the client has picked up their vehicle are sending a survey to evaluate the quality of your services, making a follow-up phone call, or asking them to please call if there are any questions or concerns. Customers appreciate having their vehicle clean and ready when they arrive to pick it up.

Task G6 Deal with angry customer/client.

Whenever you are dealing with a customer who is angry, try to determine the source of exactly why they are unhappy and investigate their complaint. If it is something that you can easily fix, you must let them know you will take care of the problem right away. If the problem is going to require some additional time to repair the vehicle, always try to arrange things in a manner in which the customer is not inconvenienced any further.

Task G7 Follow up; keep customer/client informed.

By keeping in contact with a customer throughout a repair, you make the customer feel that they are informed at every step of the repair process. Additionally, the contact lets the customer know that you have investigated any questions that they may have had about the repairs. This is also a good opportunity to ask questions about indirect damage or pre-existing damage.

Task G8 Recognize basic claims handling procedures; explain to customer/client.

When dealing with a customer, you should always explain the manner in which their claim will be processed. This should include things like the claims adjusting process, deductibles, betterment charges, and parts procurement. This will help a customer feel that they better understand the process that is followed in order to get their vehicle repaired. This is very important information to share with a customer who has never been in an accident and may not understand the process that must be followed.

Task G9 Project positive attitude and professional appearance.

By projecting a positive attitude and appearance, you will help the customer understand that they are dealing with a professional and that their repairs will be handled in the best manner possible.

Task G10 Inform customer/client about parts and the repair process.

When performing repairs on a vehicle, it is important to explain, in detail, the parts that will be repaired and the parts that will be replaced. Also, you should explain the process in which you will perform the repairs on their vehicle. This gives the customer the information to help them better understand the manner in which you will be repairing their vehicle.

Task G11 Provide warranty information.

When performing a repair on a vehicle, always explain the warranty provided on workmanship, paint, and parts. Each paint manufacturer has a warranty that will cover any paint-related failures, as do parts suppliers. Your shop has quality standards for craftsmanship, and this needs to be given to the customer in writing at the time of delivery. Also any applicable warranties for sublet operations should be provided. This will enable the customer to determine the warranty coverage for each area repaired on their vehicle.

Task G12 Provide technical and consumer protection information.

When servicing an area of the car which could affect a safety feature of the vehicle, such as an antilock brake system (ABS), or air bags/seat belts, or steering and suspension-related repairs, you should always inform the customer of the repairs that were made and the extent of the damage to those systems.

Task G13 Estimate and explain duration of out-of-service time.

When a customer is leaving their vehicle at your facility, it is important to explain the details of the repair and the estimated time that it will take to repair the damage. By explaining all the steps of the repair, the customer will be able to better understand why it will take the amount of time that you have estimated.

Task G14 Apply negotiation skills and obtain a mutual agreement.

When negotiating a price with an insurance adjuster, you must apply negotiation skills for things like repair versus replacement of a damaged area, OEM versus aftermarket parts, betterment charges, direct and indirect damage, pre-existing damage, and frame damage repair estimates.

Task G15 Interpret and explain manual or computer-assisted estimate to customer/client.

Whether an estimate is generated manually or by a computer-based estimating system, it is essential to explain your estimate to the customer. When reviewing an estimate with a customer, it is essential that you explain the details in layman's terms that are clear and easy to understand. Some of the most important areas to cover are the ones that will affect timing, quality, or cost to the customer. Some of those items could be betterment charges, deductibles, aftermarket parts, or used parts. It is of paramount importance that the customer be made aware of these items at the time of the estimate.

Sample Test for Practice

Sample Test

Please note the letter and number in parentheses following each question. They match the overview in section 4 that discusses the relevant subject matter. You may want to refer to the overview using this cross-referencing key to help with questions posing problems for you.

1. All of the following are commonly sublet operations **EXCEPT:**
   A. Towing
   B. A/C recharge
   C. Fender replacement
   D. Air bag replacement (B15)
2. Partial panel replacement, not utilizing factory seams is called:
   A. Sectioning
   B. Surfacing
   C. Cutout
   D. Splitting (B20)
3. Estimator A says that judgment time includes refinish time. Estimator B says that judgment time is listed in the collision estimating guide. Who is right?
   A. A only
   B. B only
   C. Both A and B
   D. Neither A nor B (B9)
4. Estimator A says that it is important to describe the charges and parts betterment with the customer at the time of the estimate. Estimator B says it is important to explain the difference between the aftermarket parts and OEM parts that will be used during the repair of their vehicle. Who is right?
   A. A only
   B. B only
   C. Both A and B
   D. Neither A nor B (G15)
5. All of the following should be cleaned off of a damaged vehicle before inspection **EXCEPT:**
   A. Wax
   B. Dirt
   C. Road salt
   D. Road grime (A2)
6. Estimator A says that bumpers are available as re-conditioned parts. Estimator B says that re-conditioned fenders are available. Who is right?
   A. A only
   B. B only
   C. Both A and B
   D. Neither A nor B (B14)

7. Estimator A says that a customer is rarely interested in the process in which you repair their vehicle. Estimator B says that explaining the various steps involved with a repair is essential for the customer to understand the timing estimate given to them. Who is right?
   A. A only
   B. B only
   C. Both A and B
   D. Neither A nor B (G13)

8. Estimator A uses a vehicle repair manual as a reference when evaluating suspension damage. Estimator B uses a vehicle repair manual as a reference when evaluating electrical damage. Who is right?
   A. A only
   B. B only
   C. Both A and B
   D. Neither A nor B (A12)

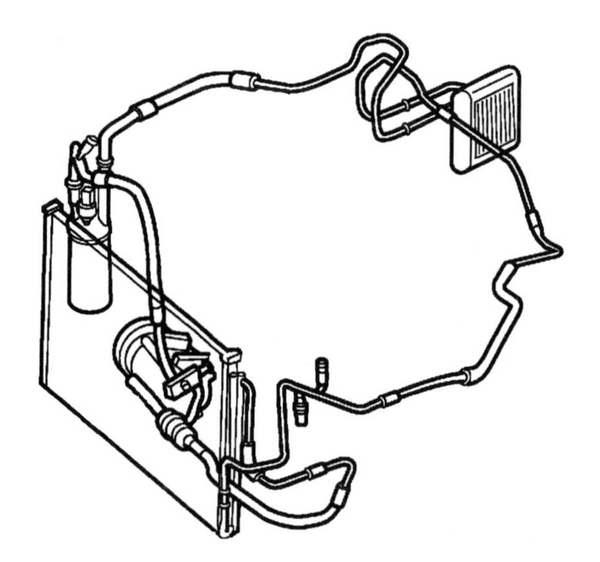

9. The air conditioning system, as shown above, has been open for five days due to parts availability. Estimator A says that the drier may also need to be replaced. Estimator B says that the evaporator needs to be replaced. Who is right?
   A. A only
   B. B only
   C. Both A and B
   D. Neither A nor B (E4.3)

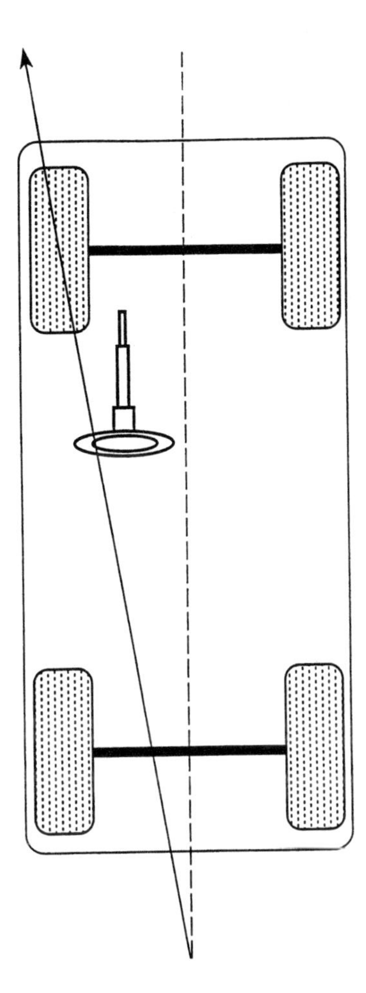

10. Estimator A says that the first step in a four-wheel alignment is the thrust angle, as shown above. Estimator B says that toe is the inward or outward tilt of the top of the wheel as viewed from the front. Who is right?
    A. A only
    B. B only
    C. Both A and B
    D. Neither A nor B (E2.3)
11. When inspecting a vehicle for damage, Estimator A uses any stall available, even if it is cluttered. Estimator B uses only a stall free of obstructions. Who is right?
    A. A only
    B. B only
    C. Both A and B
    D. Neither A nor B (A1)
12. Estimator A says that you should always inform the customer of the repairs that you performed to the vehicle, especially when it applies to a safety area of the car. Estimator B says that informing the customer of any safety related repairs should be the responsibility of the insurance company. Who is right?
    A. A only
    B. B only
    C. Both A and B
    D. Neither A nor B (G12)

13. Estimator A says that a front clip includes the radiator support. Estimator B says that a front clip includes the hood. Who is right?
    A. A only
    B. B only
    C. Both A and B
    D. Neither A nor B (F3.2)

14. All of the following are parts of an air bag system **EXCEPT:**
    A. Driver side module
    B. Crash sensors
    C. Passenger side module
    D. Motor and rail (E6.1)

15. Which of the following is not included in a manually written collision estimate?
    A. Part prices
    B. Part numbers
    C. Frame dimensions
    D. Labor allowances (B25)

16. Estimator A says that a rebuilt part should function the same as OEM. Estimator B says that a remanufactured part should not compromise the quality of the repair. Who is right?
    A. A only
    B. B only
    C. Both A and B
    D. Neither A nor B (F4.2)

17. Estimator A says that when a vehicle with faded paint is damaged, the insurance company may deduct betterment from the cost of refinish. Estimator B says that insurance companies apply betterment to only a few vehicle components. Who is right?
    A. A only
    B. B only
    C. Both A and B
    D. Neither A nor B (B27)

18. Estimator A says that a modified vehicle requires extra labor to repair. Estimator B says that accessories should be noted on the estimate. Who is right?
    A. A only
    B. B only
    C. Both A and B
    D. Neither A nor B (B3)

19. Estimator A says that by explaining the claims process to a customer, they will better understand the process that must be followed. Estimator B says that the claim process should only be explained to the customer by their insurance agent. Who is right?
    A. A only
    B. B only
    C. Both A and B
    D. Neither A nor B (G8)

20. Interior damage includes all of the following **EXCEPT:**
    A. Dash
    B. Trunk
    C. Seat
    D. Fender (A14)

21. Estimator A says that a suspension system can function properly with a bent strut. Estimator B says that a bent strut must be replaced. Who is right?
    A. A only
    B. B only
    C. Both A and B
    D. Neither A nor B (E2.2)

22. All of the following are considered when a vehicle is declared a total loss **EXCEPT:**
    A. Salvage value
    B. Retail value
    C. Repair cost
    D. Vehicle age (A6)

23. Because a bumper is made of several parts, it is called a(n):
    A. Component
    B. Structure
    C. Assembly
    D. Formulation (A10)

24. Estimator A says that add-on accessories include running boards. Estimator B says that add-on accessories include fog lights. Who is right?
    A. A only
    B. B only
    C. Both A and B
    D. Neither A nor B (D8)

25. The labor deduction taken for replacement of panels with a common seam is _______.
    A. Overlap
    B. Overhaul
    C. Overhead
    D. Overland (B17)

26. The letters EPA stand for:
    A. Environmental Protection Administration
    B. Environmental Protection Agency
    C. Environmental Projection Agency
    D. Environmental Projection Administration (C1)

27. Estimator A says that you should always identify which panels are to be repaired and which panels are to be replaced and inform the customer, prior to performing the repair. Estimator B says that as long as the quality of the repair meets the warranty requirements, it will not matter which panels are replaced and which ones are to be repaired. Who is right?
    A. A only
    B. B only
    C. Both A and B
    D. Neither A nor B (G10)

28. Estimator A says that collision estimating guides list OEM part numbers. Estimator B says that collision estimating guides list OEM part prices. Who is right?
    A. A only
    B. B only
    C. Both A and B
    D. Neither A nor B (B11)

29. Estimator A says that torque-to-yield fasteners may only be used once. Estimator B says that torque-to-yield bolts may be reused if the estimator is careful. Who is right?
    A. A only
    B. B only
    C. Both A and B
    D. Neither A nor B (E7.1)
30. The "P" pages are also known as:
    A. Process pages
    B. Procedure pages
    C. Production pages
    D. Proceed pages (B7)
31. All of the following are examples of frame measuring equipment **EXCEPT:**
    A. Laser
    B. Gauge
    C. Clamp
    D. Bench (B21)
32. All of the following vehicle aspects are used to select labor values **EXCEPT:**
    A. Vehicle color
    B. Vehicle year
    C. Vehicle body style
    D. Vehicle options (B10)
33. Estimator A says that the electrical system includes the tail lights. Estimator B says that the electrical system includes the headlights. Who is right?
    A. A only
    B. B only
    C. Both A and B
    D. Neither A nor B (E5.1)
34. Estimator A says that overlap is included in panel replacement labor allowances. Estimator B says that overlap must be added to adjacent panels. Who is right?
    A. A only
    B. B only
    C. Both A and B
    D. Neither A nor B (B17)
35. Estimator A says that windshield R&R mans to remove the windshield and repair it. Estimator B says that windshield R&R means to remove the windshield and replace it. Who is right?
    A. A only
    B. B only
    C. Both A and B
    D. Neither A nor B (D7)
36. Estimator A says that safety-related components should be replaced with OEM parts. Estimator B says that fenders are available from the OEM. Who is right?
    A. A only
    B. B only
    C. Both A and B
    D. Neither A nor B (F1.1)
37. All of the following are parts of a disk brake **EXCEPT:**
    A. Pad
    B. Shoe
    C. Caliper
    D. Rotor (E3.1)

38. To determine refinish material cost, Estimator A multiplies the refinish hours times a set dollar amount. Estimator B multiplies body labor time by a set dollar amount to determine refinish material cost. Who is right?
    A. A only
    B. B only
    C. Both A and B
    D. Neither A nor B (B19)
39. Estimator A says that if a customer is angry, you should listen to their concerns, then tell the customer whatever they want to hear so they will feel satisfied when leaving the shop, then call when they have time to think reasonably about the situation. Estimator B says that if a customer is unhappy, you should always try to reach a reasonable solution to the problem immediately. Who is right?
    A. A only
    B. B only
    C. Both A and B
    D. Neither A nor B (G6)
40. All of the following information should be obtained from the customer **EXCEPT:**
    A. Name
    B. Work phone number
    C. Work address
    D. Home address (B1)
41. Estimator A says that faulty electrical components are usually rebuilt by the body shop. Estimator B says that faulty electrical components are usually replaced by the body shop. Who is right?
    A. A only
    B. B only
    C. Both A and B
    D. Neither A nor B (E5.3)
42. Corrosion protection labor allowance needs to be figured on all of the following panels **EXCEPT:**
    A. Replacement RRIM fender
    B. Replacement steel door shell
    C. Spliced lower frame rail
    D. Spliced quarter panel (B26)
43. Estimator A says that PCM may be located in various positions either inside or outside the passenger compartment. Estimator B says that the vapor canister may be located near the fuel tank on some newer vehicles. Who is right?
    A. A only
    B. B only
    C. Both A and B
    D. Neither A nor B (E1.1)
44. Estimator A says that you should always tell the customer that you appreciate their business when they are leaving with their vehicle. Estimator B says that a follow-up phone call is a good method of contact to thank customers for their business. Who is right?
    A. A only
    B. B only
    C. Both A and B
    D. Neither A nor B (G5)

45. Estimator A says by calling a customer while you are repairing their car can cause confusion for the customer. Estimator B says that customer contact is essential when investigating any pre-existing conditions on the vehicle. Who is right?
    A. A only
    B. B only
    C. Both A and B
    D. Neither A nor B (G7)

46. The letters R&R stand for:
    A. Remove and replace
    B. Repair and replace
    C. Replace and repair
    D. Remove and repair (B4)

47. Estimator A says that a deployed air bag can be re-packed. Estimator B says that all vehicle air bag sensors must be replaced after a collision. Who is right?
    A. A only
    B. B only
    C. Both A and B
    D. Neither A nor B (A13)

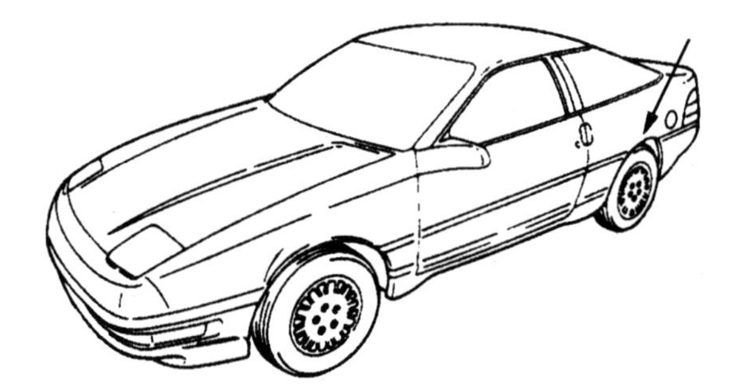

48. The part indicated is the:
    A. Front quarter panel
    B. Left fender
    C. Rear quarter panel
    D. Right fender (F2.1)

49. A heating system consists of all of the following **EXCEPT:**
    A. Core
    B. Hoses
    C. Evaporator
    D. Controls (E4.1)

50. Estimator A says that ultra high-strength steel can be repaired. Estimator B says that kinked high-strength steel can be repaired. Who is right?
    A. A only
    B. B only
    C. Both A and B
    D. Neither A nor B (D4)

51. Additional material charges would include the cost of the following **EXCEPT:**
    A. Decals
    B. Solvent
    C. Fasteners
    D. Rust proofing (B18)
52. Vehicle paint may consist of any of the following, **EXCEPT:**
    A. Single stage
    B. Basecoat/clearcoat
    C. Tri-coat
    D. Basecoat only (A11)
53. Estimator A says that all information regarding the accident should be taken from the insurance adjuster's notes. Estimator B says that the most accurate source of information regarding the claim would be from the police report. Who is right?
    A. A only
    B. B only
    C. Both A and B
    D. Neither A nor B (G2)
54. Estimator A says that the point of impact is indirect damage. Estimator B says the point of impact is direct damage. Who is right?
    A. A only
    B. B only
    C. Both A and B
    D. Neither A nor B (A4)
55. All of the following are examples of vehicle construction type **EXCEPT:**
    A. Full frame
    B. Unibody
    C. Unit frame
    D. Space frame (D1)
56. By approaching a customer with a cooperative attitude, you will accomplish all of the following **EXCEPT:**
    A. Let the customer know that they are dealing with a professional establishment
    B. Develop a trusting relationship
    C. Interfere with the insurance company's policies
    D. Improve customer satisfaction (G3)
57. Estimator A uses the collision estimating guide to determine the estimating sequence. Estimator B says that engine parts are not listed in a collision estimating guide. Who is right?
    A. A only
    B. B only
    C. Both A and B
    D. Neither A nor B (B6)
58. Estimator A says that salvage yards may have a computerized inventory. Estimator B says that salvage yards will check on the availability of parts. Who is right?
    A. A only
    B. B only
    C. Both A and B
    D. Neither A nor B (F3.4)
59. Estimator A says that back-ordered parts have the same availability as discontinued parts. Estimator B says that discontinued parts are not available. Who is right?
    A. A only
    B. B only
    C. Both A and B
    D. Neither A nor B (F1.4)

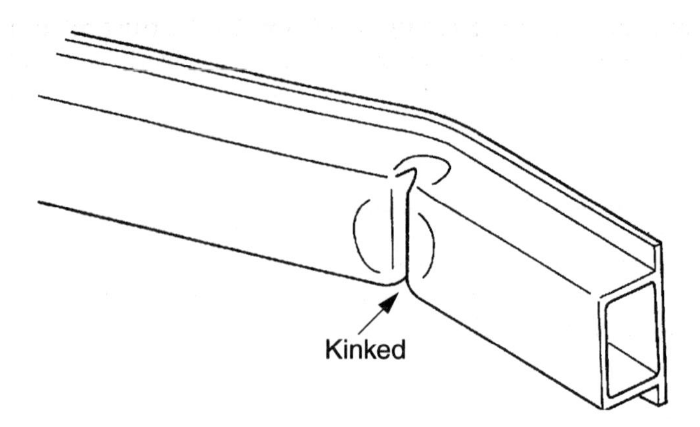

60. Estimator A says that the structural member pictured above should be repaired. Estimator B says that the structural member pictured above should be replaced. Who is right?
    A. A only
    B. B only
    C. Both A and B
    D. Neither A nor B (A7)
61. Estimator A says that seat belts should be checked after a collision. Estimator B says that seat belts do not need to be inspected after a crash. Who is right?
    A. A only
    B. B only
    C. Both A and B
    D. Neither A nor B (E6.3)

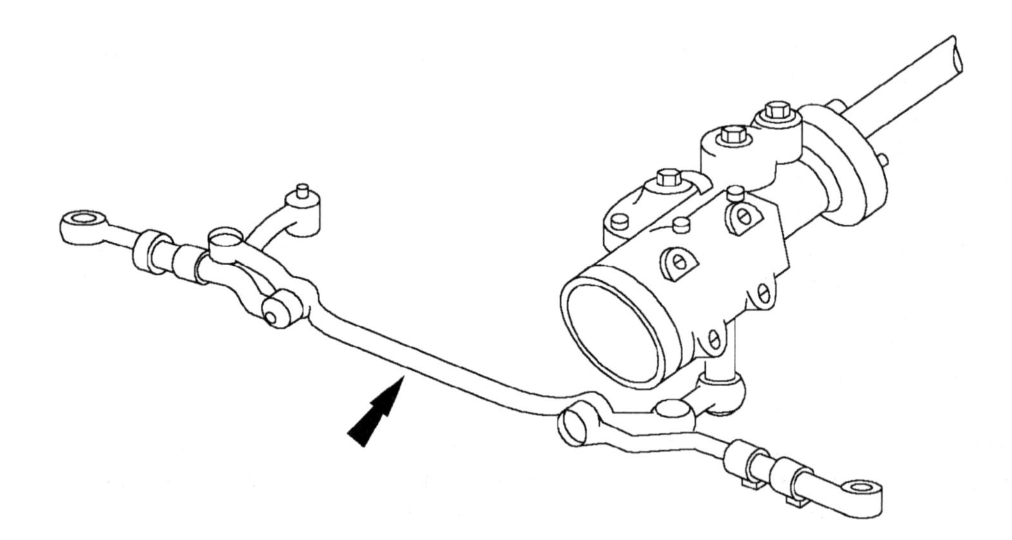

62. On a vehicle with a recirculating ball steering system, the component that is between the two inner tie rods as shown above is called:
    A. A rack and pinion
    B. An idler arm
    C. A manual steering gear
    D. A center link (E2.1)

63. Estimator A says that a warm greeting will improve the trust relationship between you and the customer. Estimator B says that a warm greeting will let the customer know that you care about their vehicle and their business. Who is right?
    A. A only
    B. B only
    C. Both A and B
    D. Neither A nor B (G1)
64. Estimator A says that add-on accessory labor times are listed in collision estimating guides. Estimator B says that add-on accessory labor times are not listed in collision estimating guides. Who is right?
    A. A only
    B. B only
    C. Both A and B
    D. Neither A nor B (A15)
65. Estimator A says that manually written estimate forms often have a column to check if a panel is to be repaired. Estimator B says that manually written estimate forms often have a column to check if a panel is to be replaced. Who is right?
    A. A only
    B. B only
    C. Both A and B
    D. Neither A nor B (B24)
66. Estimator A says that the collision shop will retain the warranty for work that is sublet. Estimator B says that you should always provide a written copy of the warranty information to the customer at the time of delivery. Who is right?
    A. A only
    B. B only
    C. Both A and B
    D. Neither A nor B (G11)
67. Estimator A says that pre-existing damage may consist of rust. Estimator B says that pre-existing damage may consist of door dings. Who is right?
    A. A only
    B. B only
    C. Both A and B
    D. Neither A nor B (A5)
68. Estimator A says that a shop should make all repairs specified on an estimate. Estimator B says that all shops offer warranties. Who is right?
    A. A only
    B. B only
    C. Both A and B
    D. Neither A nor B (C2)
69. Estimator A says that a vehicle must be moving at least 5 m.p.h. and crash to cause the air bag to inflate. Estimator B says that a continually glowing air bag light indicates the system is functional. Who is right?
    A. A only
    B. B only
    C. Both A and B
    D. Neither A nor B (E6.2)
70. Non-OEM parts are also known as:
    A. Aftermade parts
    B. Aftermarket parts
    C. Free market parts
    D. Free made parts (B12)

71. Estimator A says that a clean, professional appearance is very important in making a customer comfortable. Estimator B says that a positive attitude is important in winning the customer's trust. Who is right?
    A. A only
    B. B only
    C. Both A and B
    D. Neither A nor B (G9)
72. All of the following are examples of frame repair equipment **EXCEPT:**
    A. Slide hammer
    B. Floor pots
    C. Bench
    D. Rack (B22)
73. The function of the condenser is to change refrigerant:
    A. From solid to liquid
    B. From liquid to solid
    C. From gas to liquid
    D. From liquid to gas (E4.2)
74. Estimator A says that the availability of rebuilt parts can be checked by looking in a catalog. Estimator B says that rebuilt part availability can be verified by calling the supplier. Who is right?
    A. A only
    B. B only
    C. Both A and B
    D. Neither A nor B (F4.4)
75. Estimator A says that panel replacement decisions are based on the extent of damage. Estimator B says that panel replacement decisions are based on the cost of repair. Who is right?
    A. A only
    B. B only
    C. Both A and B
    D. Neither A nor B (F1.3)
76. Estimator A says that used parts may be specified for older vehicles. Estimator B says that the use of salvage parts should be based on the age of the damaged vehicle. Who is right?
    A. A only
    B. B only
    C. Both A and B
    D. Neither A nor B (F3.3)
77. Estimator A says that you should always review each repair with the adjuster to ensure that you are both of the same opinion as to how the repair should be performed. Estimator B says that the estimate provided by the insurance company is always the most effective and appropriate method in which to repair the vehicle and a review of the damage with the adjuster is unnecessary. Who is right?
    A. A only
    B. B only
    C. Both A and B
    D. Neither A nor B (G14)
78. Estimator A says that plastic on a vehicle may be rigid. Estimator B says that a vehicle may have flexible plastic. Who is right?
    A. A only
    B. B only
    C. Both A and B
    D. Neither A nor B (D6)

79. Estimator A says that repaired vehicles must be returned to pre-loss condition. Estimator B says that pre-loss condition includes frame dimensions. Who is right?
    A. A only
    B. B only
    C. Both A and B
    D. Neither A nor B (C3)

80. Estimator A says that aftermarket parts cost less than OEM parts. Estimator B says that the estimator must consider if aftermarket parts will return the vehicle to pre-loss conditions. Who is right?
    A. A only
    B. B only
    C. Both A and B
    D. Neither A nor B (F2.3)

81. A bumper shock should be replaced if any of the following are present **EXCEPT:**
    A. Leaking tube
    B. Bent tube
    C. Minor bend in the mounting plate
    D. Broken tube (D3)

82. Estimator A says that the alternator converts mechanical energy into electrical energy. Estimator B says that the alternator converts electrical energy into mechanical energy. Who is right?
    A. A only
    B. B only
    C. Both A and B
    D. Neither A nor B (E5.2)

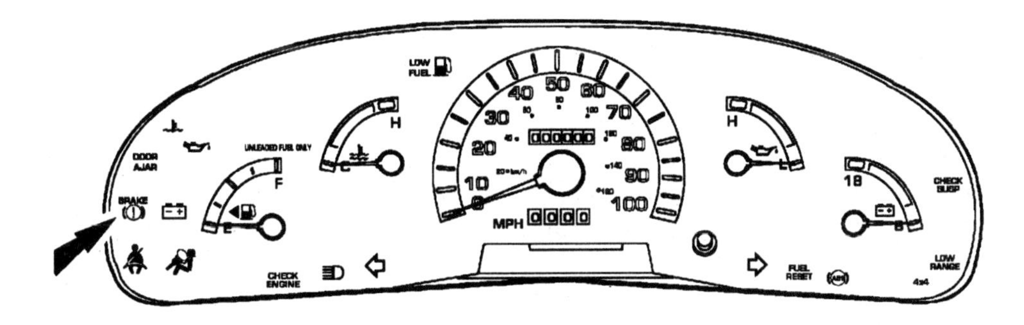

83. As shown in the figure above, a vehicle returns with the red brake light on continuously after a collision repair where the ABS hydraulic control unit was replaced. Estimator A says that the hydraulic control unit is not likely the problem, however, they would check the system. Estimator B says that it is likely not the hydraulic control unit because most hydraulic control units have nothing to do with the red brake warning indicator, and the customer might want to take the vehicle to a brake repair facility. Who is right?
    A. A only
    B. B only
    C. Both A and B
    D. Neither A nor B (E3.2)

84. The total labor hours for this estimate with 4.0 hours body labor plus 2.5 refinish hours plus 1.0 clearcoat time would be:
    A. 6.5 hours
    B. 7.0 hours
    C. 7.5 hours
    D. 8.0 hours (B23)

85. Estimator A says that the "P" pages list panel refinish times. Estimator B says that the head notes list refinish time. Who is right?
A. A only
B. B only
C. Both A and B
D. Neither A nor B (B8)

86. Estimator A says that body repair materials include body filler. Estimator B says that refinishing materials include clearcoat. Who is right?
A. A only
B. B only
C. Both A and B
D. Neither A nor B (E7.3)

87. Estimator A says that the left side of a vehicle is the driver's side. Estimator B says that the left and right sides of a vehicle are determined as if you are sitting in the vehicle. Who is right?
A. A only
B. B only
C. Both A and B
D. Neither A nor B (B5)

88. A unibody vehicle that may be repaired may have any of the following frame damage **EXCEPT:**
A. Diamond
B. Twist
C. Sag
D. Mash (D2)

89. Estimator A says that used parts cost the same as OEM parts. Estimator B says that used parts are usually sold as assemblies. Who is right?
A. A only
B. B only
C. Both A and B
D. Neither A nor B (B13)

90. Estimator A says that rebuilt parts fit the same as OEM. Estimator B says that remanufactured parts may have a warranty. Who is right?
A. A only
B. B only
C. Both A and B
D. Neither A nor B (F4.3)

91. Estimator A says that an aftermarket fender may not have all of the necessary holes pre-drilled. Estimator B says that in order to be used , an aftermarket part must function the same as an OEM part. Who is right?
A. A only
B. B only
C. Both A and B
D. Neither A nor B (F2.2)

92. All of the following are examples of salvage assemblies **EXCEPT:**
A. Rear clip
B. Bumper cover
C. Rear clip top
D. Front clip (F3.1)

93. Estimator A uses a tram gauge to measure center line. Estimator B uses a tram gauge to measure height. Who is right?
   A. A only
   B. B only
   C. Both A and B
   D. Neither A nor B (A8)

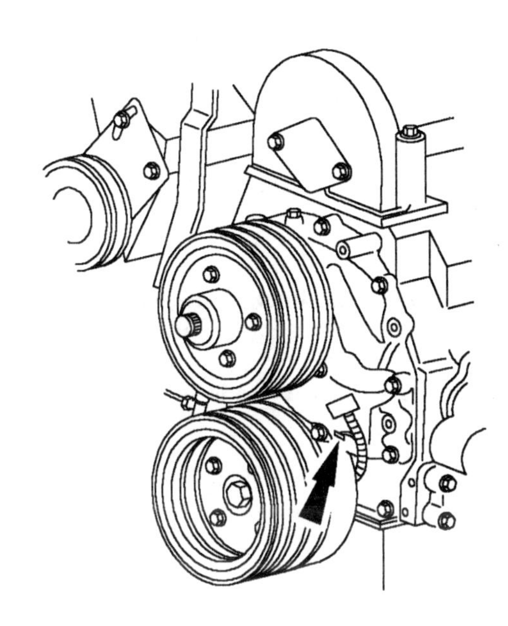

94. A vehicle with a transverse mounted engine has a wire pulled loose from a sensor (shown above) behind the harmonic balancer. Estimator A says that could be the reason for the no spark/no fuel situation. Estimator B says that the knock sensor will not cause a no start. Who is right?
   A. A only
   B. B only
   C. Both A and B
   D. Neither A nor B (E1.2)
95. Which of the following must be considered when evaluating the quality of repair?
   A. Integrity
   B. Cost
   C. Speed
   D. Location (A3)
96. Estimator A says that aluminum may be welded with a specially equipped MIG welder. Estimator B says that aluminum may be welded with a TIG welder. Who is right?
   A. A only
   B. B only
   C. Both A and B
   D. Neither A nor B (D5)
97. Estimator A says that even though an aftermarket part is listed in a catalog, it may not be available. Estimator B says that the availability of aftermarket parts can be verified by contacting the OEM dealer. Who is right?
   A. A only
   B. B only
   C. Both A and B
   D. Neither A nor B (F2.4)

98. Estimator A says that by identifying yourself on the initial phone contact from the customer you give the customer the impression that you are a true professional. Estimator B says that allowing the customer to contact you when they have questions will build a trusting relationship. Who is right?
    A. A only
    B. B only
    C. Both A and B
    D. Neither A nor B (G4)

99. A door has slight, two-hour estimated repair time, damage. Estimator A says that we should replace the door shell. Estimator B says that we should replace the door skin. Who is right?
    A. A only
    B. B only
    C. Both A and B
    D. Neither A nor B (A9)

100. A vehicle with rear drum brakes has been damaged in the right rear wheel area. The brake backing plate is bent and must be replaced. Estimator A says that the brake shoes must be replaced. Estimator B says that an allowance must be added to the estimate to cover brake bleeding. Who is right?
    A. A only
    B. B only
    C. Both A and B
    D. Neither A nor B (E3.3)

101. The letters VIN stand for:
    A. Vehicle information number
    B. Vehicle information numeral
    C. Vehicle identification number
    D. Vehicle identification numeral (B2)

102. Estimator A says that structural parts support the weight of the vehicle. Estimator B says that structural parts absorb collision impact. Who is right?
    A. A only
    B. B only
    C. Both A and B
    D. Neither A nor B (F1.2)

103. When inspecting a bolt to determine if it is metric or standard, on a metric bolt you would expect to see:
    A. Slotted marks on the head
    B. No marks on the head
    C. Dots on the head
    D. A number on the head (E7.2)

104. Estimator A says that a disconnected vacuum line causes a rough idle on newer vehicles. Estimator B says that a disconnected vacuum line causes a high idle on older vehicles. Who is right?
    A. A only
    B. B only
    C. Both A and B
    D. Neither A nor B (E1.3)

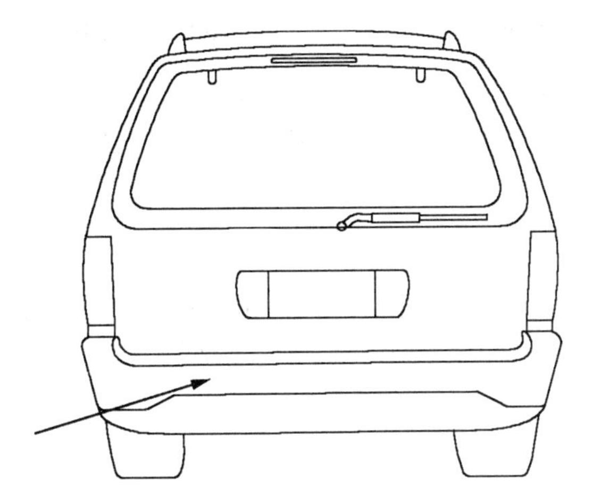

105. The remanufactured part pictured above is a:
    A. Bumper
    B. Bumper cover
    C. Bumper reinforcement
    D. Impact absorber (F4.1)

6 Additional Test Questions for Practice

Additional Test Questions

Please note the letter and number in parentheses following each question. They match the overview in section 4 that discusses the relevant subject matter. You may want to refer to the overview using this cross-referencing key to help with questions posing problems for you.

1. By listening to the customer with your full attention, you will accomplish all of the following **EXCEPT** to:
   A. let the customer know that they are dealing with a professional establishment.
   B. develop a trusting relationship.
   C. improve customer satisfaction.
   D. irritate other customers. (G3)
2. All of the following replacement parts need to be refinished **EXCEPT:**
   A. Hoods
   B. Bumper reinforcements
   C. Fenders
   D. Lower frame rails (A11)
3. Estimator A says that included operations are part of another labor operation. Estimator B says that alignment is included in overhaul. Who is right?
   A. A only
   B. B only
   C. Both A and B
   D. Neither A nor B (B4)
4. Estimator A says that used parts may have rust damage. Estimator B says that used parts may have previous repaired damage. Who is right?
   A. A only
   B. B only
   C. Both A and B
   D. Neither A nor B (B13)
5. All of the following may be needed to inspect a damaged vehicle **EXCEPT:**
   A. Lift
   B. Trouble light
   C. Computer estimating system
   D. Clean stall (A1)
6. A parallelogram steering system may include all of the following parts **EXCEPT:**
   A. tie rod.
   B. rack and pinion.
   C. center link.
   D. drag link. (E2.1)

7. The power window on a damaged door does not operate. Estimator A says that the wiring may be broken. Estimator B says that the regulator may be damaged. Who is right?
   A. A only
   B. B only
   C. Both A and B
   D. Neither A nor B (E5.3)
8. A customer has a question about how his vehicle will be repaired. Estimator A tells the customer to read the estimate. Estimator B takes the time to explain the repair process. Who is right?
   A. A only
   B. B only
   C. Both A and B
   D. Neither A nor B (G10)
9. Estimator A sends a survey to the customer after repairs are made. Estimator B calls each customer after repairs are made. Who is right?
   A. A only
   B. B only
   C. Both A and B
   D. Neither A nor B (G5)
10. Estimator A always greets a customer in a friendly manner. Estimator B is indifferent to greeting customers. Who is right?
    A. A only
    B. B only
    C. Both A and B
    D. Neither A nor B (G1)
11. All of the following are electrical system components **EXCEPT:**
    A. manual door locks.
    B. power door locks.
    C. batteries.
    D. head lights. (E5.1)
12. Estimator A says that the insurance company pays for betterment. Estimator B says that the vehicle's owner pays for betterment. Who is right?
    A. A only
    B. B only
    C. Both A and B
    D. Neither A nor B (A5)
13. Estimator A says that set-up time includes time to anchor the vehicle. Estimator B says that the set-up time includes time to position the vehicle. Who is right?
    A. A only
    B. B only
    C. Both A and B
    D. Neither A nor B (B22)
14. Estimator A says that when a customer brings their vehicle to you, it shows that they already have trust in your abilities to repair their vehicle. Estimator B says that if you make it known that you appreciate their trust and that you truly understand their situation, it will make them feel good that they have chosen a professional such as yourself. Who is right?
    A. A only
    B. B only
    C. Both A and B
    D. Neither A nor B (G1)

15. A reference book will be used to determine refinish material cost. Estimator A says that the paint code is used. Estimator B says that the refinish hours are used. Who is right?
    A. A only
    B. B only
    C. Both A and B
    D. Neither A nor B (B19)
16. A vapor canister may be located in any of the following locations **EXCEPT:**
    A. In the glove box
    B. Near the gas tank
    C. On the apron
    D. In the engine compartment (E1.1)
17. Windshield R&I means that the windshield was removed and:
    A. Repaired
    B. Replaced
    C. Put back
    D. Discarded (D7)
18. The letters R&I stand for:
    A. Remove and improve
    B. Repair and improve
    C. Replace and install
    D. Remove and install (B4)
19. An A/C system recharge would include adding:
    A. oil and water to the system.
    B. oil and refrigerant to the system.
    C. water and refrigerant to the system.
    D. water and oil to the system. (E4.3)
20. Estimator A says that the estimator determines judgment time. Estimator B says that slightly damaged panels are usually repaired rather than replaced. Who is right?
    A. A only
    B. B only
    C. Both A and B
    D. Neither A nor B (B9)
21. Estimator A says that a remanufactured bumper cover costs more than OEM. Estimator B says that a remanufactured bumper cover costs less than OEM. Who is right?
    A. A only
    B. B only
    C. Both A and B
    D. Neither A nor B (F4.3)
22. Estimator A says that interchangeable parts, such as a fender, means that a fender for one vehicle may be the same as the fender for another type of vehicle. Estimator B says that the bumper fascia, bumper reinforcement and impact absorber are collectively called the bumper assembly. Who is right?
    A. A only
    B. B only
    C. Both A and B
    D. Neither A nor B (B5)

23. Estimator A says that even though aftermarket part numbers are listed, the part may not be available. Estimator B says that if the aftermarket part is listed in a catalog, it will be available. Who is right?
    A. A only
    B. B only
    C. Both A and B
    D. Neither A nor B (B12)
24. All of the following are available as aftermarket parts **EXCEPT:**
    A. Lower frame rail
    B. Fender
    C. Bumper
    D. Hood (B12)
25. Estimator A says that sectioning must be made in approved locations only. Estimator B says that sectioning saves labor time when compared to full replacement. Who is right?
    A. A only
    B. B only
    C. Both A and B
    D. Neither A nor B (B20)
26. Estimator A records vehicle body style when writing an estimate. Estimator B records the paint code. Who is right?
    A. A only
    B. B only
    C. Both A and B
    D. Neither A nor B (B2)
27. All of the following are not included in fender replacement **EXCEPT:**
    A. Bumper removal
    B. Hole drilling
    C. Refinishing
    D. Side marker lamp replacement (B7)

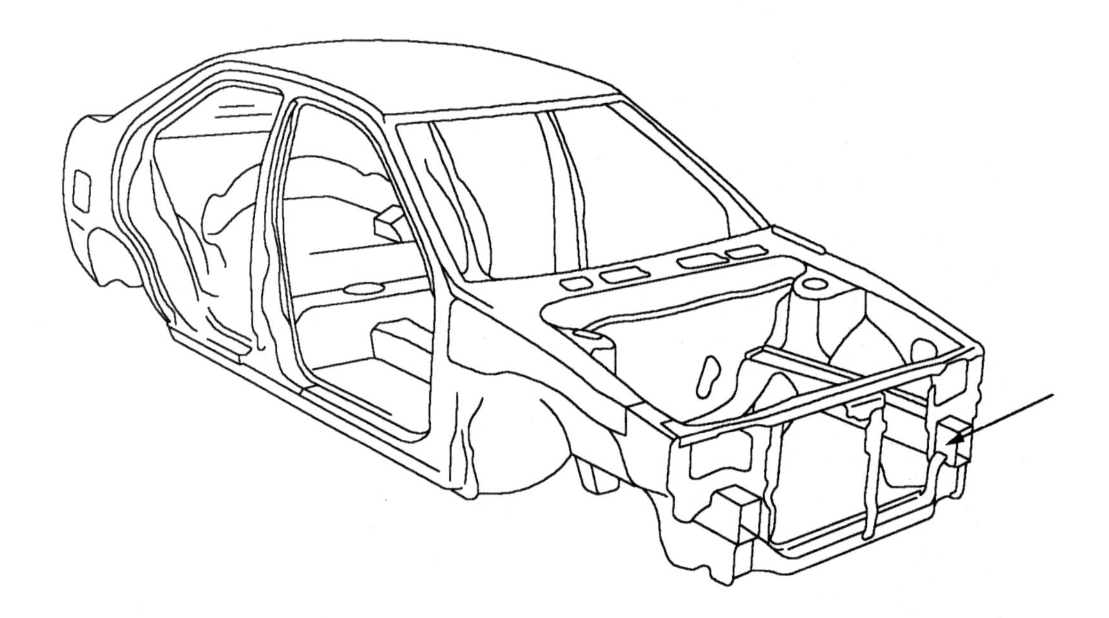

28. The part indicated is the:
    A. apron.
    B. lower frame rail.
    C. strut assembly.
    D. strut tower. (F1.1)

29. The power train for a front wheel drive vehicle includes all of the following **EXCEPT:**
    A. Transaxle
    B. Engine
    C. Half shafts
    D. Transmission (E2.1)
30. In a collision estimating guide, the assemblies for each model are arranged:
    A. Back to front
    B. Back to side
    C. Front to back
    D. Side to front (B6)
31. Estimator A says that in a space frame construction, the outer body panels contribute to the strength of the vehicle. Estimator B says that in a space frame construction, the body panels do not contribute to the strength of the vehicle. Who is right?
    A. A only
    B. B only
    C. Both A and B
    D. Neither A nor B (D1)
32. Estimator A uses the paint code and color book to determine if a white vehicle has clearcoat. Estimator B says that the vehicle should be inspected for signs of a previous repaint. Who is right?
    A. A only
    B. B only
    C. Both A and B
    D. Neither A nor B (A11)
33. Estimator A says that commonly needed collision parts may be in stock at the OEM dealer. Estimator B says that the OEM dealer will supply part availability information. Who is right?
    A. A only
    B. B only
    C. Both A and B
    D. Neither A nor B (F1.4)
34. Estimator A says that labor allowance is dependent on vehicle body style. Estimator B says that vehicle make determines labor allowance. Who is right?
    A. A only
    B. B only
    C. Both A and B
    D. Neither A nor B (B10)
35. An air bag has been deployed. Estimator A says the air bag light will glow. Estimator B says the air bag light will not glow. Who is right?
    A. A only
    B. B only
    C. Both A and B
    D. Neither A nor B (E6.2)
36. Which of the following panels need a corrosion protection labor allowance:
    A. Urethane bumper
    B. SMC door
    C. Steel fender
    D. RRIM fender (B26)

37. Estimator A says that OSHA requires that workers are informed about hazardous chemicals. Estimator B says that EPA regulates workers' health. Who is right?
    A. A only
    B. B only
    C. Both A and B
    D. Neither A nor B (C1)

38. The term used to describe improving a damaged vehicle's condition to better than pre-accident condition is called:
    A. An upgrade
    B. Depreciation
    C. Stabilization
    D. Betterment (A5)

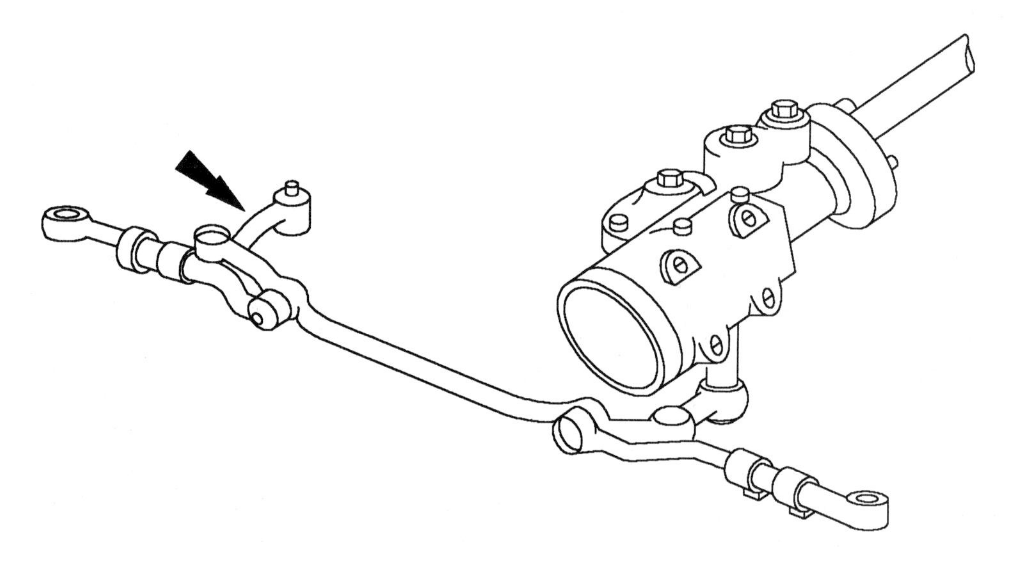

39. Estimator A says an alignment is necessary when replacing an idler arm as shown above. Estimator B says that a good steering technician will not need to perform an alignment when replacing a single tie rod end. Who is right?
    A. A only
    B. B only
    C. Both A and B
    D. Neither A nor B (E2.3)

40. You should always inform the customer of the repairs that were made and the extent of the damage to the:
    A. body panels.
    B. safety systems.
    C. lighting systems.
    D. underbody. (G12)

41. Which of the following must always be replaced after air bag deployment?
    A. Crash sensors
    B. Clock spring
    C. Dash panel
    D. Air bag module (E6.3)
42. Estimator A says that frame measurement time includes time to diagnose the damage. Estimator B says that frame measurement time includes time to pull the frame. Who is right?
    A. A only
    B. B only
    C. Both A and B
    D. Neither A nor B (B22)
43. All of the following are included in a knee **EXCEPT:**
    A. strut.
    B. knuckle.
    C. control arm.
    D. half shaft. (F3.1)
44. The safety system that places the shoulder belt on the seat occupant is known as:
    A. manual seat belt.
    B. driver's side air bag.
    C. automatic seat belt.
    D. passenger's side air bag. (E6.1)
45. Estimator A says that body repair supplies include grinder disks. Estimator B says that refinishing supplies include sand paper. Who is right?
    A. A only
    B. B only
    C. Both A and B
    D. Neither A nor B (E7.3)
46. Estimator A welds aluminum with an arc welder. Estimator B uses a TIG welder to weld aluminum. Who is right?
    A. A only
    B. B only
    C. Both A and B
    D. Neither A nor B (D5)
47. Estimator A says that the use of a rebuilt part should not compromise the quality of the repair. Estimator B says that a reconditioned part should function the same as OEM. Who is right?
    A. A only
    B. B only
    C. Both A and B
    D. Neither A nor B (F4.2)

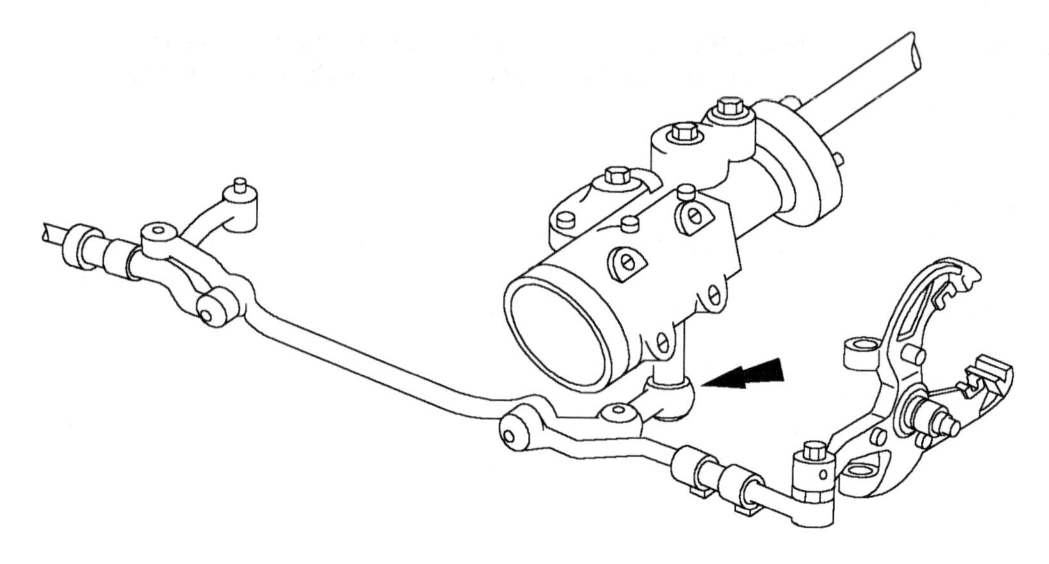

48. The component shown above is:
    A. a connector link.
    B. an idler arm.
    C. a pitman arm.
    D. a connecting rod. (E2.1)
49. Estimator A says that a properly sectioned vehicle is as crash worthy as an undamaged vehicle. Estimator B says that a properly sectioned vehicle is not as durable as an undamaged vehicle. Who is right?
    A. A only
    B. B only
    C. Both A and B
    D. Neither A nor B (B21)
50. Estimator A says that foot notes override the information in the "P" pages. Estimator B says that head notes include R&I time. Who is right?
    A. A only
    B. B only
    C. Both A and B
    D. Neither A nor B (B8)
51. Estimator A marks up the sublet fee by 50%. Estimator B marks up the sublet fee by 75%. Who is right?
    A. A only
    B. B only
    C. Both A and B
    D. Neither A nor B (B15)
52. Estimator A says that the brake warning light will glow if the brake system loses pressure. Estimator B says a glowing brake warning light indicates an ABS malfunction. Who is right?
    A. A only
    B. B only
    C. Both A and B
    D. Neither A nor B (E3.2)
53. Overhaul includes all of the following **EXCEPT:**
    A. Remove the part from the vehicle
    B. Install the part on the vehicle
    C. Refinish part off of the vehicle
    D. Inspect part off of the vehicle (B4)

54. A part is backordered. Estimator A says that the part may be available in a week. Estimator B says that the part may be available in a month. Who is right?
    A. A only
    B. B only
    C. Both A and B
    D. Neither A nor B (F1.4)

55. Estimator A says that a vacuum leak on newer vehicles will not cause any driveability problems. Estimator B says that fuel injector cleaner should be added to the fuel tank at every fill-up. Who is right?
    A. A only
    B. B only
    C. Both A and B
    D. Neither A nor B (E1.3)

56. A drum brake consists of all of the following **EXCEPT:**
    A. Rotor
    B. Drum
    C. Shoe
    D. Wheel cylinder (E3.1)

57. Calling the customer provides all of the following **EXCEPT:**
    A. a chance to answer questions.
    B. the opportunity to ask about pre-existing damage.
    C. a chance to explain the repair.
    D. the opportunity to complain about the insurance company. (G7)

58. Estimator A says that a statement of quality and workmanship from your facility should be provided at the time of delivery to the customer. Estimator B says that you should always provide a written copy of the warranty information to the customer at the time of delivery. Who is right?
    A. A only
    B. B only
    C. Both A and B
    D. Neither A nor B (G11)

59. Estimator A says that the type of fastener used in replacement is important. Estimator B says that any bolt that will fit is all right. Who is right?
    A. A only
    B. B only
    C. Both A and B
    D. Neither A nor B (E7.1)

60. Estimator A says you should explain the processes that are to be used in the repairs to the customer. Estimator B says that you should always identify which panels are to be repaired and which panels will be replaced with the customer prior to performing the repair. Who is right?
    A. A only
    B. B only
    C. Both A and B
    D. Neither A nor B (G10)

61. Estimator A says that a broken styrofoam impact absorber can be taped back together and reused. Estimator B says that a dent in a styrofoam impact absorber should be filled with plastic body filler. Who is right?
    A. A only
    B. B only
    C. Both A and B
    D. Neither A nor B (D3)

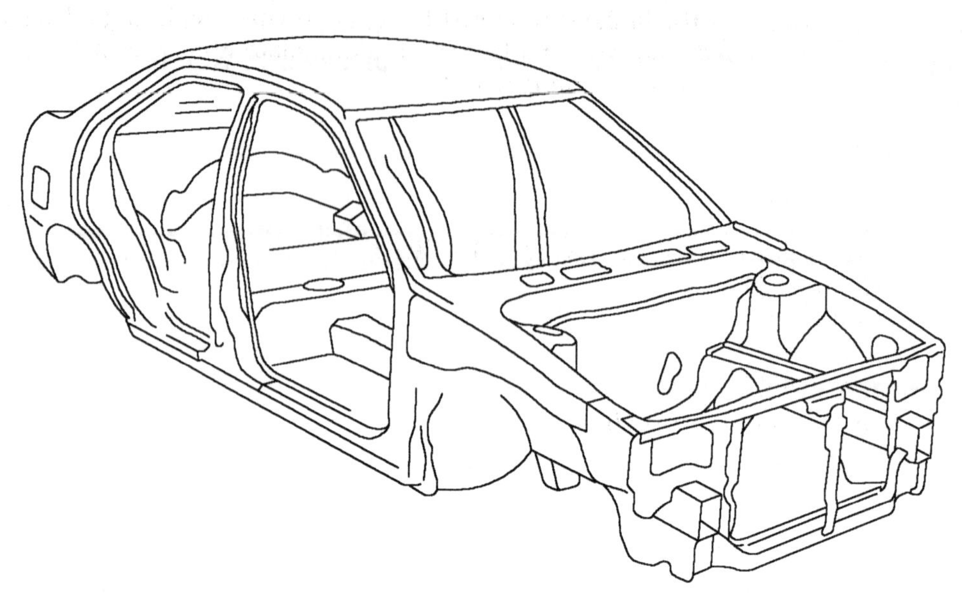

62. The figure above shows a:
    A. space frame.
    B. unibody.
    C. full frame.
    D. partial frame. (D1)
63. Estimator A says the customer has no interest in the repair of his vehicle, so he never contacts the customer during repair. Estimator B calls the customer as the repair progresses. Who is right?
    A. A only
    B. B only
    C. Both A and B
    D. Neither A nor B (G7)
64. Estimator A uses a tram to measure diagonal. Estimator B uses a tram to measure length. Who is right?
    A. A only
    B. B only
    C. Both A and B
    D. Neither A nor B (A8)
65. Estimator A listens to the customer describe the vehicle damage. Estimator B ignores the customer's input because he is the collision repair professional and the customer is not. Who is right?
    A. A only
    B. B only
    C. Both A and B
    D. Neither A nor B (G2)
66. Estimator A says that a used fender assembly includes the fender liner. Estimator B says a used fender assembly includes moldings. Who is right?
    A. A only
    B. B only
    C. Both A and B
    D. Neither A nor B (F3.1)

67. Estimator A says that the brake shoes must be replaced when replacing a backing plate. Estimator B says that when replacing a backing plate, the shoes may be reused, if they are not worn. Who is right?
   A. A only
   B. B only
   C. Both A and B
   D. Neither A nor B (E3.3)

68. All of the following would be recorded **EXCEPT:**
   A. Production date
   B. Gas mileage
   C. Mileage
   D. License plate number (B2)

69. The most commonly used OEM parts are:
   A. sheet metal.
   B. alternators.
   C. the hood.
   D. safety related parts. (F1.1)

70. Estimator A says that depreciation, based on mileage, may be taken on damaged suspension parts. Estimator B says that betterment may be deducted for existing faded paint. Who is right?
   A. A only
   B. B only
   C. Both A and B
   D. Neither A nor B (B27)

71. All of the following are safety related components **EXCEPT:**
   A. Air bags
   B. ABS
   C. Tie rod ends
   D. Fender (G12)

72. The letters VOC stand for:
   A. Various organic compounds
   B. Variable organic compounds
   C. Various organic chemicals
   D. Volatile organic compounds (C1)

73. Estimator A says that R-12 and R-134a can be used interchangeably. Estimator B says that R-12 and R-134a cannot be used interchangeably. Who is right?
   A. A only
   B. B only
   C. Both A and B
   D. Neither A nor B (E4.2)

74. Estimator A says that a unibody is designed to absorb collision energy. Estimator B says that convolutions in unibody structural members absorb impact. Who is right?
   A. A only
   B. B only
   C. Both A and B
   D. Neither A nor B (D2)

75. Estimator A says that a deployed air bag must be replaced. Estimator B says that all sensors must be replaced after deployment. Who is right?
   A. A only
   B. B only
   C. Both A and B
   D. Neither A nor B (A13)

76. All of the following are included in a rear clip **EXCEPT:**
    A. rear doors.
    B. deck lid.
    C. tail lights.
    D. rear bumper. (F3.2)
77. What is used to determine the price of a vehicle?
    A. Percentage of new price
    B. Owner's manual
    C. Used car value guide
    D. Collision estimating guide (A6)
78. Estimator A says that only unibody vehicles can have mash damage. Estimator B says that only full frame vehicles can have mash damage. Who is right?
    A. A only
    B. B only
    C. Both A and B
    D. Neither A nor B (D2)
79. All of the following are body repair material **EXCEPT:**
    A. Body filler
    B. MIG weld wire
    C. Clearcoat
    D. Plastic repair material (E7.3)
80. All of the following are warranted by the shop **EXCEPT:**
    A. Preexisting damage
    B. Parts
    C. Paint
    D. Workmanship (G11)
81. Estimator A says that brake bleeding is required whenever a brake system is opened. Estimator B says that a good technician can replace a brake line without bleeding. Who is right?
    A. A only
    B. B only
    C. Both A and B
    D. Neither A nor B (E3.3)
82. When writing an estimate, Estimator A records the customer information first. When writing an estimate, Estimator B checks the vehicle paint code first. Who is right?
    A. A only
    B. B only
    C. Both A and B
    D. Neither A nor B (B1)
83. Estimator A says the battery stores electrical energy. Estimator B says that a broken wire is often the cause of electrical component malfunction. Who is right?
    A. A only
    B. B only
    C. Both A and B
    D. Neither A nor B (E5.2)

84. Estimator A says that frame measurement labor time should be included in all estimates. Estimator B says that frame measurement labor time should be only included in estimates of vehicles with frame damage. Who is right?
    A. A only
    B. B only
    C. Both A and B
    D. Neither A nor B (B22)

85. The front suspension is jounced to test for:
    A. Steering gear box damage
    B. Front wheel alignment
    C. Steering rack alignment
    D. Steering arm damage (A12)

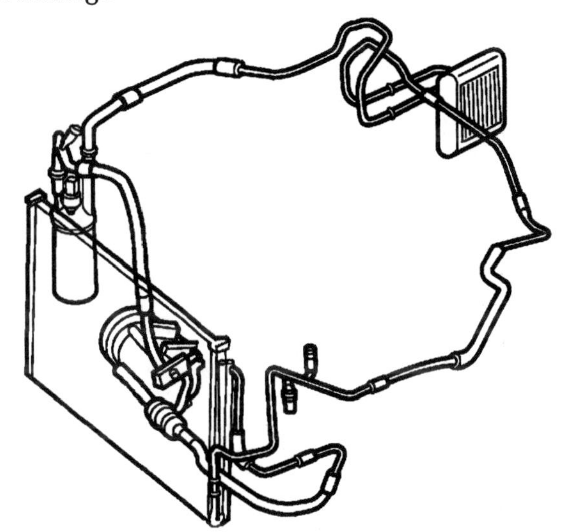

86. The figure above shows:
    A. a fuel pressure system.
    B. a heater system.
    C. an A/C system.
    D. a fuel evaporative emission system. (E4.1)

87. Estimator A says that encapsulated glass requires special care in removal. Estimator B says that movable glass may be bolted to the window regulator. Who is right?
    A. A only
    B. B only
    C. Both A and B
    D. Neither A nor B (D7)

88. Estimator A says that broken wires may be repaired by soldering. Estimator B says that broken wires may be repaired by crimping. Who is right?
    A. A only
    B. B only
    C. Both A and B
    D. Neither A nor B (E5.3)

89. Estimator A says that used parts are sometimes installed on newer vehicles. Estimator B says that the decision to use salvage parts should be based on vehicle condition. Who is right?
    A. A only
    B. B only
    C. Both A and B
    D. Neither A nor B (F3.3)

90. Estimator A says that if repair cost exceeds the retail value, a damaged vehicle is a total loss. Estimator B says that total loss vehicle's salvage value is variable. Who is right?
    A. A only
    B. B only
    C. Both A and B
    D. Neither A nor B (A6)
91. Estimator A says that a positive attitude is important in winning the trust of the customer. Estimator B says that if your attitude is positive, and it is obvious that you run a busy shop, the appearance of a messy desk is not so important. Who is right?
    A. A only
    B. B only
    C. Both A and B
    D. Neither A nor B (G9)
92. Estimator A answers the shop phone by saying "Hello". Estimator B answers the shop phone by saying the shop's name and his own name. Who is right?
    A. A only
    B. B only
    C. Both A and B
    D. Neither A nor B (G4)
93. Estimator A always opens the hood, even if it is crushed, to inspect for under hood damage on a front hit vehicle. Estimator B says that opening the hood is unnecessary on a light front hit. Who is right?
    A. A only
    B. B only
    C. Both A and B
    D. Neither A nor B (A2)
94. Centering the steering wheel includes all of the following **EXCEPT:**
    A. Unloading the suspension
    B. Turning the steering wheel to lock
    C. Counting the number of turns
    D. Dividing the number of turns by 2 (A12)
95. The usual markup for a used part is:
    A. 5–10%
    B. 10–15%
    C. 15–20%
    D. 20–25% (B13)
96. The refinish overlap deduction is 0.4 hours for each adjacent panel. A refinish job consisting of four panels would have how much of an overlap deduction?
    A. 0.4
    B. 0.8
    C. 1.2
    D. 1.6 (B23)
97. Estimator A says that the OEM part prices change often. Estimator B says that for the estimate to be accurate, the data must be current. Who is right?
    A. A only
    B. B only
    C. Both A and B
    D. Neither A nor B (B24)

98. Estimator A says that add-on accessories include decals/stripes. Estimator B says that add-on accessories include bed liners. Who is right?
    A. A only
    B. B only
    C. Both A and B
    D. Neither A nor B (D8)

99. Which of the following is the customer LEAST likely to know:
    A. Insurance company
    B. Work phone number
    C. Home address
    D. Claim number (B1)

100. A remanufactured bumper is also called a:
    A. retread bumper.
    B. rechromed bumper.
    C. remade bumper.
    D. redone bumper. (F4.3)

101. Estimator A says that "O" rings should be replaced when A/C lines are disconnected. Estimator B says that if an A/C system is under pressure, an A/C line can be disconnected to release the refrigerant. Who is right?
    A. A only
    B. B only
    C. Both A and B
    D. Neither A nor B (E4.3)

102. Estimator A says that frayed seat belts should be replaced. Estimator B says that over-stressed seat belts should be replaced. Who is right?
    A. A only
    B. B only
    C. Both A and B
    D. Neither A nor B (A14)

103. Estimator A says that an accumulator may also be called a drier. Estimator B says that the evaporator is also called a drier. Who is right?
    A. A only
    B. B only
    C. Both A and B
    D. Neither A nor B (E4.1)

104. Estimator A says that a fender should be removed to check for cowl damage on a cowl hit. Estimator B says that the hood should be removed to check for hidden damage on a front hit unibody. Who is right?
    A. A only
    B. B only
    C. Both A and B
    D. Neither A nor B (A2)

105. Which of the following parts would be most likely to be available aftermarket:
    A. Quarter panel
    B. Apron
    C. Center pillar
    D. Fender (F2.4)

106. A kink is described as a distortion:
    A. greater than 90 degrees over a short radius.
    B. less than 90 degrees over a short radius.
    C. greater than 60 degrees over a long radius.
    D. less than 60 degrees over a short radius. (A7)

107. Estimator A says that all warranties are for one year maximum. Estimator B says all warranties are for however long the customer owns the vehicle. Who is right?
   A. A only
   B. B only
   C. Both A and B
   D. Neither A nor B (C2)
108. Estimator A says an air bag system includes crash sensors. Estimator B says that an air bag system includes a module. Who is right?
   A. A only
   B. B only
   C. Both A and B
   D. Neither A nor B (E6.1)
109. Which of the following is not included in collision estimating guides:
   A. Add-on accessories guides
   B. OEM part labor time
   C. OEM part refinish time
   D. OEM part overhaul time (A15)
110. Estimator A says that tire depreciation is based on remaining tread. Estimator B says that suspension components are subject to depreciation. Who is right?
   A. A only
   B. B only
   C. Both A and B
   D. Neither A nor B (B27)
111. Estimator A says that the shop must be able to answer the customer's questions. Estimator B says that the customer will trust a shop that explains all of the procedures. Who is right?
   A. A only
   B. B only
   C. Both A and B
   D. Neither A nor B (G3)
112. Estimator A says that options do not need to be listed on every estimate. Estimator B says that the presence of safety systems, such as air bags, should be listed on the estimate. Who is right?
   A. A only
   B. B only
   C. Both A and B
   D. Neither A nor B (B3)
113. Estimator A says that the air bag system goes through self checks each time the key is turned on. Estimator B says that in some types of crashes, the air bag or bags do not deploy. Who is right?
   A. A only
   B. B only
   C. Both A and B
   D. Neither A nor B (E6.2)
114. Estimator A says that according to the "P" pages, additional labor must be added to fender replacement for front bumper removal. Estimator B says that "P" pages list what is not included in labor allowances. Who is right?
   A. A only
   B. B only
   C. Both A and B
   D. Neither A nor B (B7)

115. When inspecting bolts and nuts to determine if they are metric or standard, to indicate American standard bolts, you would expect to see:
   A. letters on the head.
   B. no marks on the head.
   C. lines or dots on the head.
   D. a number on the head. (E7.2)
116. All of the following may be damaged by objects inside the vehicle except the:
   A. Deck lid
   B. Quarter panel
   C. Fender
   D. Door (A14)
117. Estimator A says that panel fit is considered a pre-loss condition. Estimator B says that color match is considered a pre-loss condition. Who is right?
   A. A only
   B. B only
   C. Both A and B
   D. Neither A nor B (C3)
118. Estimator A says that not all parts that may be needed are available as used. Estimator B says that used late model vehicle parts may be difficult to locate. Who is right?
   A. A only
   B. B only
   C. Both A and B
   D. Neither A nor B (F3.4)
119. Ultra high strength steel is used to make:
   A. Door crash beams
   B. Mild steel
   C. High strength steel
   D. Quarter panels (D4)

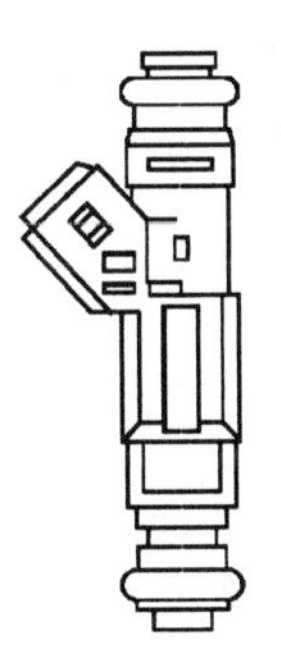

120. The component shown above is:
   A. an A/C pressure relief valve.
   B. an emission sensor.
   C. a fuel pressure regulator.
   D. a fuel injector. (E1.1)
121. Estimator A says that a small amount of play is acceptable if it is within tolerances, in some steering components. Estimator B says that suspension parts are subject to wear. Who is right?
   A. A only
   B. B only
   C. Both A and B
   D. Neither A nor B (E2.2)

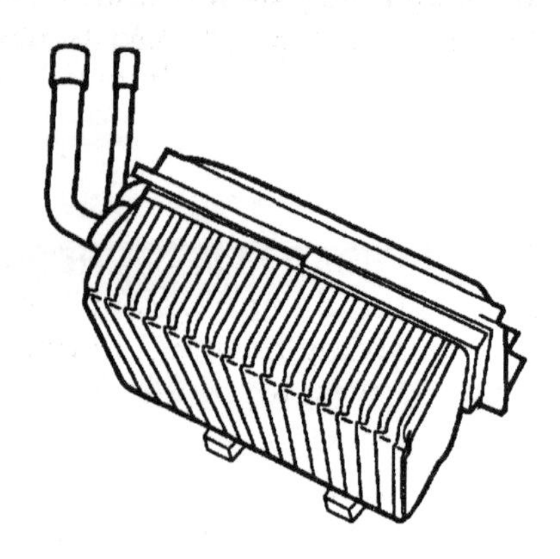

122. The air conditioning evaporator shown above:
    A. is located behind the radiator.
    B. is located in front of the radiator.
    C. changes refrigerant gas to liquid form.
    D. is located within the heater case. (E4.1)
123. Estimator A says that if the ABS warning light glows, the system is operational. Estimator B says that a glowing ABS warning light indicates a system malfunction. Who is right?
    A. A only
    B. B only
    C. Both A and B
    D. Neither A nor B (E3.2)
124. The name of the part that connects the steering gear box to the center link is called:
    A. the drag link.
    B. the pitman arm.
    C. the tie rod.
    D. the idler arm. (E2.1)
125. Estimator A says that all damaged parts must be restored to pre-loss condition. Estimator B says that paint durability is considered a pre-loss condition. Who is right?
    A. A only
    B. B only
    C. Both A and B
    D. Neither A nor B (C3)
126. Estimator A says that unibody vehicles have a separate frame. Estimator B says that a space frame vehicle is similar to a full frame vehicle. Who is right?
    A. A only
    B. B only
    C. Both A and B
    D. Neither A nor B (D1)
127. Estimator A says that the set-up time for a floor pot system is the same as the set-up time for a bench. Estimator B says that a bench is used to repair light damage only. Who is right?
    A. A only
    B. B only
    C. Both A and B
    D. Neither A nor B (B22)

128. Estimator A says that a shop may do less than the amount of work on an estimate, if the customer agrees. Estimator B says that no deletions from the estimate are allowed unless the customer authorizes them. Who is right?
    A. A only
    B. B only
    C. Both A and B
    D. Neither A nor B (C2)

129. When determining the point of impact, Estimator A says that the vehicle should be inspected. When determining the point of impact, Estimator B says that the customer should be interviewed, if possible. Who is right?
    A. A only
    B. B only
    C. Both A and B
    D. Neither A nor B (A4)

130. The part indicated is the:
    A. rocker panel.
    B. a pillar.
    C. center pillar.
    D. quarter panel. (F1.1)

131. Estimator A says that slightly damaged suspension components can be straightened. Estimator B says that the estimator must determine the most cost effective method to restore the damaged vehicle to pre-accident condition. Who is right?
    A. A only
    B. B only
    C. Both A and B
    D. Neither A nor B (F1.3)

132. All of the following are commonly available aftermarket parts **EXCEPT:**
    A. Fender
    B. Bumper
    C. Header panel
    D. Frame rail (F2.4)

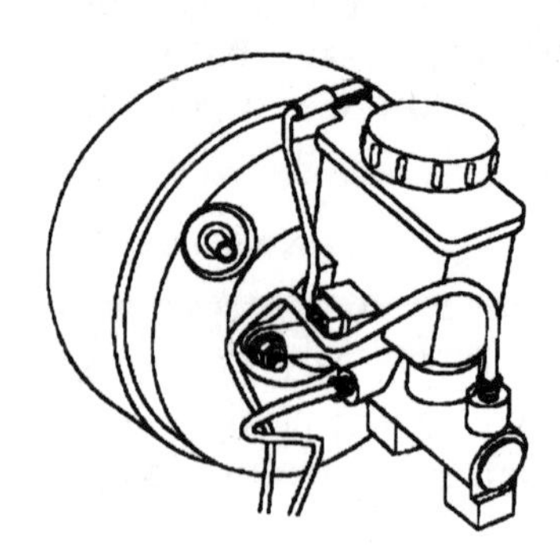

133. The component shown above is:
   A. an anti-lock brake hydraulic unit.
   B. a power steering pump assembly.
   C. a master cylinder and power booster assembly.
   D. a manual brake assembly. (E3.1)
134. Which of the following, if not protected, will corrode the fastest?
   A. Urethane bumper cover
   B. Spliced frame rail
   C. Steel fender
   D. SMC hatch (B26)
135. Estimator A says that stabilizer bars may be found on the front suspension on some vehicles. Estimator B says that stabilizer bars may be found on the rear suspension of some vehicles. Who is right?
   A. A only
   B. B only
   C. Both A and B
   D. Neither A nor B (E2.1)
136. Estimator A says that an aftermarket part will always have all of the necessary holes pre-drilled. Estimator B says that in some cases, aftermarket fenders need holes drilled. Who is right?
   A. A only
   B. B only
   C. Both A and B
   D. Neither A nor B (F2.2)
137. Which of the following electrical components is least likely to be damaged in a frontal collision?
   A. Head light
   B. Starter
   C. Turn signal lamp
   D. Battery (E5.2)
138. Estimator A says that calling a customer while you are repairing their vehicle can create customer confidence in your shop. Estimator B says that insurance company contact is essential to investigate any pre-existing conditions on the vehicle. Who is right?
   A. A only
   B. B only
   C. Both A and B
   D. Neither A nor B (G7)

139. This vehicle part is:
    A. right quarter panel.
    B. left quarter panel.
    C. right fender.
    D. left fender. (F1.1)

140. Estimator A says that it is important to explain the advantages of a computer-assisted estimate to the customer. Estimator B says that it is important to explain the complete estimate to the customer. Who is right?
    A. A only
    B. B only
    C. Both A and B
    D. Neither A nor B (G15)

141. Estimator A says the customer should be informed as to when the repaired vehicle will be finished. Estimator B tells the customer not to call, the shop will call when the vehicle is finished. Who is right?
    A. A only
    B. B only
    C. Both A and B
    D. Neither A nor B (G13)

142. Estimator A says that overlap must be figured into the refinish labor times. Estimator B says that included labor times must be added to the estimate. Who is right?
    A. A only
    B. B only
    C. Both A and B
    D. Neither A nor B (B17)

143. Which of the following should be part of the additional materials charge?
    A. Stripe tape
    B. Body filler
    C. Body side molding
    D. Decals (B18)

144. If the refinish hours are 7.5 and the shop's per hour refinish materials charge is $20/hour, the refinish material's amount would be:
    A. 150.00
    B. 160.00
    C. 170.00
    D. 180.00 (B19)

145. Which of the following is not good business practice when the customer is picking up their vehicle?
   A. send a survey to evaluate your services.
   B. make a follow-up phone call.
   C. ask them to please call if there are any questions or concerns.
   D. offer to clean their vehicle when they show up. (G5)
146. All of the following can be used to measure the center line **EXCEPT:**
   A. Laser
   B. Tram gauge
   C. Universal bench
   D. Self-centering gauges (A8)
147. Estimator A repairs a flexible plastic bumper fascia with body filler. Estimator B repairs a flexible plastic bumper fascia with adhesive. Who is right?
   A. Estimator A
   B. Estimator B
   C. Both A and B
   D. Neither A nor B (A9)
148. By not identifying yourself on the initial phone contact with a prospective customer you:
   A. start a trusting relationship.
   B. instill confidence in your business.
   C. instill confidence that they are working with a true professional.
   D. create a question of your honesty in the mind of the client. (G4)
149. Estimator A says that standard bolts have lines or dots on the head. Estimator B says that metric bolts have dots on the head. Who is right?
   A. A only
   B. B only
   C. Both A and B
   D. Neither A nor B (E7.2)
150. Estimator A says that sublet fees may be less than standard fees. Estimator B says that bumper replacement is commonly sublet. Who is right?
   A. A only
   B. B only
   C. Both A and B
   D. Neither A nor B (B15)
151. Estimator A says that computer-generated estimates list part prices. Estimator B says that computer-generated estimates list labor allowances. Who is right?
   A. A only
   B. B only
   C. Both A and B
   D. Neither A nor B (B24)
152. When inspecting a vehicle for damage, Estimator A starts at the point of impact. When inspecting a vehicle for damage, Estimator B starts with the point farthest from the damage. Who is right?
   A. A only
   B. B only
   C. Both A and B
   D. Neither A nor B (A10)

153. A customer is unhappy with a repair. Estimator A tells the customer "We did the best that we could". Estimator B tells the customer that he will look at the problem to see if the complaint is justified. Who is right?
    A. A only
    B. B only
    C. Both A and B
    D. Neither A nor B (G6)
154. Estimator A says that most vehicle parts are available as remanufactured. Estimator B says that the supplier should be contacted to verify remanufactured part availability. Who is right?
    A. A only
    B. B only
    C. Both A and B
    D. Neither A nor B (F4.4)
155. Estimator A says that suspension fasteners may be torque-to-yield. Estimator B always checks the manufacturer's recommendations to see if fasteners are torque-to-yield. Who is right?
    A. A only
    B. B only
    C. Both A and B
    D. Neither A nor B (E7.1)
156. Estimator A says that minor bends in a bumper shock mounting plate can be repaired. Estimator B says that if a bumper shock has even minor damage at the mounting plate, it has lost its strength and must be replaced. Who is right?
    A. A only
    B. B only
    C. Both A and B
    D. Neither A nor B (D3)
157. Estimator A says that included labor is listed in the "P" pages. Estimator B says that included labor is listed in each section, if it applies. Who is right?
    A. A only
    B. B only
    C. Both A and B
    D. Neither A nor B (B17)
158. All of the following are non-structural parts **EXCEPT:**
    A. Hood
    B. Rocker panel
    C. Door
    D. Quarter panel (A9)
159. All of the following are found in an air conditioning system **EXCEPT:**
    A. Evaporator
    B. Condenser
    C. Expansion valve
    D. Heater core (E4.1)
160. Estimator A says that the right rear wheel is on the passenger's side. Estimator B says that the right rear wheel is on the driver's side. Who is right?
    A. A only
    B. B only
    C. Both A and B
    D. Neither A nor B (B5)

161. Estimator A says that body filler may be used to repair a dented aluminum panel. Estimator B says that on some vehicles, the hood is made of aluminum. Who is right?
 A. A only
 B. B only
 C. Both A and B
 D. Neither A nor B (D5)
162. The amount of inertia damage is influenced by the following **EXCEPT:**
 A. Vehicle weight
 B. Vehicle height
 C. Vehicle speed
 D. Vehicle occupant's weight (A14)
163. The letters SUV stand for:
 A. Special utility vehicle
 B. Special use vehicle
 C. Sport utility vehicle
 D. Sport utilization vehicle (D1)
164. If a labor operation is needed but it is not an included item, it is called a(n):
 A. Add-on
 B. Overlap
 C. Addition
 D. Deduction (B17)
165. Estimator A says a two-channel ABS system controls the rear wheels together. Estimator B says a two-channel ABS controls the rear wheels independently. Who is right?
 A. A only
 B. B only
 C. Both A and B
 D. Neither A nor B (E3.1)
166. The most common type of steel on a vehicle is:
 A. Ultra high-strength steel
 B. Mild steel
 C. High strength steel
 D. Martensetic steel (D4)
167. Estimator A says that damaged steering components should be straightened. Estimator B says that bent suspension components should be straightened. Who is right?
 A. A only
 B. B only
 C. Both A and B
 D. Neither A nor B (E2.3)
168. Estimator A says that add-on accessories are included in collision estimating guides. Estimator B says that collision estimating guides do not include add-on accessories. Who is right?
 A. A only
 B. B only
 C. Both A and B
 D. Neither A nor B (D8)
169. Safety components protect the:
 A. passengers.
 B. drive train.
 C. engine.
 D. transmission. (F1.2)

170. Estimator A says that the cost of bolts needs to be part of additional materials in an estimate. Estimator B says that name plates need to be included in additional materials. Who is right?
    A. A only
    B. B only
    C. Both A and B
    D. Neither A nor B (B18)
171. Estimator A says that all imported vehicles are listed in one guide. Estimator B uses a collision estimating guide to determine the estimating sequence. Who is right?
    A. A only
    B. B only
    C. Both A and B
    D. Neither A nor B (B6)
172. Estimator A says that a special adhesion promoter must be used to repair olefin plastics. Estimator B says that rigid plastic may be repaired with adhesives. Who is right?
    A. A only
    B. B only
    C. Both A and B
    D. Neither A nor B (D6)
173. Reconditioned metal bumpers are called:
    A. Redone bumpers
    B. Remade bumpers
    C. Rechromed bumpers
    D. Refinished bumpers (B14)
174. Estimator A says that flexible plastics may be reshaped with heat. Estimator B says that all types of plastic may be welded. Who is right?
    A. A only
    B. B only
    C. Both A and B
    D. Neither A nor B (D6)
175. The part number is listed as 66635-6. Estimator A says that 66635 is the number of the left hand part. Estimator B says that 66635 is the number of the right hand part. Who is right?
    A. A only
    B. B only
    C. Both A and B
    D. Neither A nor B (B11)
176. Estimator A says that the labor allowance for applying corrosion protection to a new, non-welded replacement part should include time to apply seam sealer. Estimator B says that a welded splice is not subject to rust. Who is right?
    A. A only
    B. B only
    C. Both A and B
    D. Neither A nor B (B26)
177. Estimator A says that the procedure used to produce a computer generated estimate varies according to the information company. Estimator B says that computer generated estimates include part numbers. Who is right?
    A. A only
    B. B only
    C. Both A and B
    D. Neither A nor B (B24)

178. Estimator A says that some damaged panels are not repairable. Estimator B says that damaged steering parts must always be replaced. Who is right?
  A. A only
  B. B only
  C. Both A and B
  D. Neither A nor B (F1.3)

179. Estimator A says that aftermarket parts have different packaging than OEM parts. Estimator B says that some sheet metal parts are available as aftermarket. Who is right?
  A. A only
  B. B only
  C. Both A and B
  D. Neither A nor B (F2.1)

180. When inspecting for damage, Estimator A looks for panel gaps. Estimator B looks for stress cracks when inspecting for damage. Who is right?
  A. A only
  B. B only
  C. Both A and B
  D. Neither A nor B (A4)

181. An estimate is being explained to a customer. Estimator A uses technical jargon. Estimator B uses layman's language. Who is right?
  A. A only
  B. B only
  C. Both A and B
  D. Neither A nor B (G15)

182. Which of the following is not included in judgment time:
  A. Rough out
  B. Refinish
  C. Analyze damage
  D. Plan repair (B9)

183. A rear clip top is to be installed on a rear hit vehicle. Estimator A says that the rocker panels will be spliced and welded. Estimator B says that the rear frame rail will be spliced and welded. Who is right?
  A. A only
  B. B only
  C. Both A and B
  D. Neither A nor B (F3.3)

184. Estimator A replaces all damaged panels on a vehicle. Estimator B replaces only panels in which the repair cost exceeds the replacement cost. Who is right?
  A. A only
  B. B only
  C. Both A and B
  D. Neither A nor B (A3)

185. Estimator A says that a unibody vehicle can have sidesway damage. Estimator B says that a full frame vehicle can have sidesway damage. Who is right?
  A. A only
  B. B only
  C. Both A and B
  D. Neither A nor B (D2)

186. Estimator A says that if a customer is angry, you should try to find the source of their unhappiness. Estimator B says that if the customer is unhappy, you should always try to reach a reasonable solution to the problem. Who is right?
   A. A only
   B. B only
   C. Both A and B
   D. Neither A nor B (G6)
187. Estimator A says that you should always review each repair with the adjuster and apply negotiation skills for things like repair versus replacement of components and aftermarket versus OEM components. Estimator B says you should always review each repair with the adjuster and discuss things like betterment charges, direct and indirect damage, and pre-existing damage. Who is right?
   A. A only
   B. B only
   C. Both A and B
   D. Neither A nor B (G14)
188. All of the following are commonly remanufactured **EXCEPT:**
   A. fenders.
   B. wheels.
   C. bumpers.
   D. alternators. (F4.1)
189. A vehicle has been in a rear-end collision and will not start. Estimator A says that the wiring to the in-tank fuel pump could be the cause. Estimator B says that the inertia switch could cause a no-start condition. Who is right?
   A. A only
   B. B only
   C. Both A and B
   D. Neither A nor B (E1.2)
190. What must be considered when determining if a damaged panel should be repaired or replaced?
   A. Placement on the vehicle
   B. Cost of the repair
   C. Color of the part
   D. Composition of the part (A3)
191. A pickup truck has aftermarket fender flares. One of the flares is damaged. Estimator A says that the labor time to replace the flare is listed in the collision estimating guide. Estimator B says that the flare price is listed in the collision estimating guide. Who is right?
   A. A only
   B. B only
   C. Both A and B
   D. Neither A nor B (A15)
192. Estimator A says that the collision estimating guide lists OEM parts only. Estimator B says that some OEM parts listed in the guide may not be available. Who is right?
   A. A only
   B. B only
   C. Both A and B
   D. Neither A nor B (B11)
193. Customers will accept what they may perceive as a long out-of-service time if you:
   A. show an understanding of insurance processes.
   B. explain all the steps of the necessary repairs.
   C. give a shorter than expected repair time.
   D. promise to call if more time is needed. (G13)

194. Estimator A says that rebuilt part availability should be confirmed. Estimator B says that alternators are available as rebuilt parts. Who is right?
    A. A only
    B. B only
    C. Both A and B
    D. Neither A nor B (B14)

195. Estimator A uses a collision estimating guide to determine the part number. Estimator B uses a collision estimating guide to determine the part location. Who is right?
    A. A only
    B. B only
    C. Both A and B
    D. Neither A nor B (A10)

196. To calculate clearcoat time, the refinish hours are multiplied by 0.4. A panel requiring 2.5 hours to refinish would require how many hours to clearcoat?
    A. 0.5
    B. 1.0
    C. 1.5
    D. 2.0 (B23)

197. Estimator A says that a positive attitude is enthusiasm for the job. Estimator B says that a positive attitude is a can-do spirit. Who is right?
    A. A only
    B. B only
    C. Both A and B
    D. Neither A nor B (G9)

198. Estimator A says that labor allowance is dependent on vehicle year. Estimator B says that allowance is dependent on vehicle options. Who is right?
    A. A only
    B. B only
    C. Both A and B
    D. Neither A nor B (B10)

199. Estimator A says that a damaged pressure line will cause a system malfunction. Estimator B says that the accumulator removes moisture from the A/C system. Who is right?
    A. A only
    B. B only
    C. Both A and B
    D. Neither A nor B (E4.2)

200. All of the following could cause a no-start **EXCEPT:**
    A. Severed electric fuel pump wiring
    B. Inertia switch
    C. Severed knock sensor wiring
    D. Disconnected spark and fuel sensor (E1.2)

201. All of the following are available as aftermarket **EXCEPT:**
    A. Alternator
    B. Air bag module
    C. Idler arm
    D. Fender (F2.1)

202. Estimator A says that the use of aftermarket parts allows more parts to be replaced instead of repaired. Estimator B says that aftermarket parts cost the same as OEM parts. Who is right?
   A. A only
   B. B only
   C. Both A and B
   D. Neither Estimator A nor Estimator B (F2.3)

203. Estimator A says that it is important to listen to customer concerns and needs in order to create trust in your shop. Estimator B says that it is important to listen to customer concerns and needs in order to determine preexisting conditions and indirect damage. Who is right?
   A. A only
   B. B only
   C. Both A and B
   D. Neither A nor B (G2)

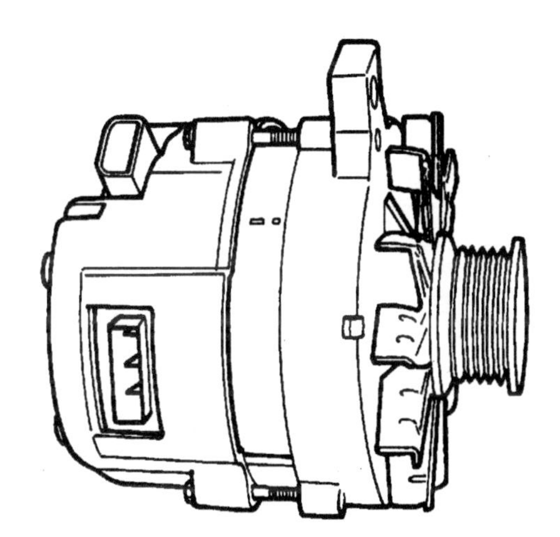

204. The part pictured above is a(n):
   A. starter.
   B. compressor.
   C. alternator.
   D. power steering pump. (F4.1)

205. What is the most important aspect to consider when a structured part is kinked?
   A. Loss of strength
   B. Loss of weight
   C. Loss of paint
   D. Loss of dimension (A7)

206. Head notes list all of the following **EXCEPT:**
   A. Refinish time
   B. Judgment time
   C. R&I time
   D. O/H time (B8)

207. Negotiation may be used on all of the following **EXCEPT:**
   A. Warranty
   B. Aftermarket parts
   C. Betterment
   D. Frame damage time (G14)

208. The seat belt should be replaced if it shows any of the following **EXCEPT:**
   A. Cut webbing
   B. Bowed webbing
   C. Stains
   D. Broken threads (E6.3)

209. Estimator A says that deployed air bags can be repacked. Estimator B says that all crash sensors must be replaced after deployment. Who is right?
   A. A only
   B. B only
   C. Both A and B
   D. Neither A nor B (E6.3)

210. Which frame type is the strongest?
   A. Full frame
   B. Unibody
   C. Space frame
   D. Unit frame (D2)

7 Appendices

Answers to the Test Questions for the Sample Test Section 5

1. C
2. A
3. D
4. C
5. A
6. A
7. B
8. C
9. A
10. A
11. B
12. A
13. C
14. D
15. C
16. C
17. A
18. C
19. A
20. D
21. B
22. D
23. C
24. C
25. A
26. B
27. A
28. C
29. A
30. B
31. C
32. A
33. C
34. D
35. B
36. C
37. B
38. A
39. B
40. C
41. B
42. A
43. C
44. C
45. B
46. A
47. D
48. B
49. C
50. D
51. A
52. D
53. D
54. B
55. C
56. C
57. A
58. C
59. B
60. B
61. A
62. D
63. C
64. B
65. C
66. B
67. C
68. A
69. D
70. B
71. C
72. A
73. C
74. C
75. C
76. C
77. A
78. C
79. C
80. C
81. C
82. A
83. A
84. C
85. B
86. C
87. C
88. A
89. B
90. C
91. C
92. B
93. D
94. A
95. A
96. C
97. A
98. C
99. D
100. B
101. C
102. C
103. D
104. D
105. B

Explanations to the Answers for the Sample Test Section 5

Question #1
Answer A is wrong. Towing may be sublet.
Answer B is wrong. A/C recharge may be sublet.
Answer C is correct. Fender replacement is usually not sublet.
Answer D is wrong. Air bag replacement may be sublet.

Question #2
Answer A is correct. Partial panel replacement is sectioning.
Answer B is wrong. Partial panel replacement is sectioning.
Answer C is wrong. Partial panel replacement is sectioning.
Answer D is wrong. Partial panel replacement is sectioning.

Question #3
Answer A is wrong. Refinish time is not part of judgment time.
Answer B is wrong. Judgment time is not listed in a collision estimating guide.
Answer C is wrong.
Answer D is correct.

Question #4
Answer A is wrong. Estimator A is right.
Answer B is wrong. Estimator B is right.
Answer C is correct.
Answer D is wrong.

Question #5
Answer A is correct. Wax does not prevent inspection.
Answer B is wrong. Dirt should be cleaned off.
Answer C is wrong. Road salt should be cleaned off.
Answer D is wrong. Road grime should be cleaned off.

Question #6
Answer A is correct. Reconditioned bumpers are available.
Answer B is wrong. Reconditioned fenders are not available.
Answer C is wrong.
Answer D is wrong.

Question #7
Answer A is wrong. Customers are usually interested in the repair process.
Answer B is correct. A customer will understand the time of repair if all steps are explained.
Answer C is wrong.
Answer D is wrong.

Question #8
Answer A is wrong. Estimator A is right.
Answer B is wrong. Estimator B is right.
Answer C is correct.
Answer D is wrong.

Question #9
Answer A is correct. If the system is open to the air for more than an hour, the drier should be replaced.
Answer B is wrong. Only if the evaporator is damaged, should it be replaced.
Answer C is wrong.
Answer D is wrong.

Question #10
Answer A is correct. The thrust angle is checked first in an alignment.
Answer B is wrong. The inward and outward tilt of the top of the wheels is camber.
Answer C is wrong.
Answer D is wrong.

Question #11
Answer A is wrong. It is difficult to adequately see damage if the work area is cluttered.
Answer B is correct. Damage analysis is easier in an area free of obstructions.
Answer C is wrong.
Answer D is wrong.

Question #12
Answer A is correct. The shop should explain all repairs, especially safety related repairs.
Answer B is wrong. The shop should explain all repairs, especially safety related repairs.
Answer C is wrong.
Answer D is wrong.

Question #13
Answer A is wrong. Estimator A is right.
Answer B is wrong. Estimator B is right.
Answer C is correct.
Answer D is wrong.

Question #14
Answer A is wrong. The driver's side module is part of the air bag system.
Answer B is wrong. Crash sensors are part of the air bag system.
Answer C is wrong. The passenger's side module is part of the air bag system.
Answer D is correct. The motor and rail part of an automatic seat belt system, not an air bag system.

Question #15
Answer A is wrong. Part numbers are provided.
Answer B is wrong. Part prices are provided.
Answer C is correct. Frame dimensions are not provided.
Answer D is wrong. Labor allowances are provided.

Question #16
Answer A is wrong. Estimator A is right.
Answer B is wrong. Estimator B is right.
Answer C is correct.
Answer D is wrong.

Question #17
Answer A is correct. Repaint of a faded vehicle will make the vehicle better than before the damage. Betterment may be deducted.
Answer B is wrong. Almost any vehicle part could have betterment.
Answer C is wrong.
Answer D is wrong.

Question #18
Answer A is wrong. Estimator A is right.
Answer B is wrong. Estimator B is right.
Answer C is correct.
Answer D is wrong.

Question #19
Answer A is correct. The shop should explain the claims process to the customer.
Answer B is wrong. Both the shop and the insurance agent can explain the claims process.
Answer C is wrong.
Answer D is wrong.

Question #20
Answer A is wrong. The dash is part of the interior.
Answer B is wrong. The trunk is part of the interior.
Answer C is wrong. The seat is part of the interior.
Answer D is correct. The fender is part of the exterior, not the interior.

Question #21
Answer A is wrong. A bent strut must be replaced.
Answer B is correct. A bent strut must be replaced.
Answer C is wrong.
Answer D is wrong.

Question #22
Answer A is wrong. Salvage value is one factor to consider.
Answer B is wrong. Retail value is one factor to consider.
Answer C is wrong. Repair cost is one factor to consider.
Answer D is correct. Vehicle age is not a factor.

Question #23
Answer A is wrong. A component is an individual part.
Answer B is wrong. A structure is an individual part.
Answer C is correct. An assembly is made of several parts.
Answer D is wrong. A formulation is not a body part.

Question #24
Answer A is wrong. Estimator A is right.
Answer B is wrong. Estimator B is right.
Answer C is correct.
Answer D is wrong.

Question #25
Answer A is correct. Overlap is the labor deduction for replacement of panels with a common seam.
Answer B is wrong. Overlap is the labor deduction for replacement of panels with a common seam.
Answer C is wrong. Overlap is the labor deduction for replacement of panels with a common seam.
Answer D is wrong. Overlap is the labor deduction for replacement of panels with a common seam.

Question #26
Answer A is wrong. The letters EPA stand for Environmental Protection Agency.
Answer B is correct. The letters EPA stand for Environmental Protection Agency.
Answer C is wrong. The letters EPA stand for Environmental Protection Agency.
Answer D is wrong. The letters EPA stand for Environmental Protection Agency.

Question #27
Answer A is correct. The customer should be informed about panel repair and replacement.
Answer B is wrong. The customer should be informed about panel repair and replacement.
Answer C is wrong.
Answer D is wrong.

Question #28
Answer A is wrong. Estimator A is right.
Answer B is wrong. Estimator B is right.
Answer C is correct.
Answer D is wrong.

Question #29
Answer A is correct. A torque-to-yield bolt must be replaced if it is removed.
Answer B is wrong. A torque-to-yield bolt must be replaced if it is removed.
Answer C is wrong.
Answer D is wrong.

Question #30
Answer A is wrong. The "P" pages are procedure pages.
Answer B is correct. The "P" pages are procedure pages.
Answer C is wrong. The "P" pages are procedure pages.
Answer D is wrong. The "P" pages are procedure pages.

Question #31
Answer A is wrong. A laser can be used to measure a frame.
Answer B is wrong. A gauge can be used to measure a frame.
Answer C is correct. A clamp cannot be used to measure a frame.
Answer D is wrong. A bench can be used to measure a frame.

Question #32
Answer A is correct. Vehicle color does not usually influence labor time.
Answer B is wrong. Vehicle year can influence labor value.
Answer C is wrong. Vehicle body style can influence labor value.
Answer D is wrong. Vehicle options can influence labor value.

Question #33
Answer A is wrong. Estimator A is right.
Answer B is wrong. Estimator B is right.
Answer C is correct.
Answer D is wrong.

Question #34
Answer A is wrong. Overlap is not included.
Answer B is wrong. Overlap must be deducted.
Answer C is wrong.
Answer D is correct.

Question #35
Answer A is wrong. Windshield R&R means to remove and replace the windshield.
Answer B is correct. Windshield R&R means to remove and replace the windshield.
Answer C is wrong. Because A is wrong, then C is wrong.
Answer D is wrong. Because B is right, then D is wrong.

Question #36
Answer A is wrong. Estimator A is right.
Answer B is wrong. Estimator B is right.
Answer C is correct.
Answer D is wrong.

Question #37
Answer A is wrong. A pad is part of a disk brake.
Answer B is correct. A shoe is not part of a disk brake.
Answer C is wrong. A caliper is part of a disk brake.
Answer D is wrong. A rotor is part of a disk brake.

Question #38
Answer A is correct. One way to calculate refinish material cost is to multiply refinish hours times a set dollar amount.
Answer B is wrong. Body labor is not used to calculate refinish material.
Answer C is wrong.
Answer D is wrong.

Question #39
Answer A is wrong. Estimator A is right.
Answer B is correct. Estimator B is right.
Answer C is wrong.
Answer D is wrong.

Question #40
Answer A is wrong. Name should be obtained.
Answer B is wrong. Work number should be obtained.
Answer C is correct. Work address usually isn't needed.
Answer D is wrong. Home address should be obtained.

Question #41
Answer A is wrong. Faulty electric components are usually replaced, not rebuilt.
Answer B is correct. Faulty electric components are usually replaced, not rebuilt.
Answer C is wrong.
Answer D is wrong.

Question #42
Answer A is correct. A plastic RRIM fender does not need corrosion protection.
Answer B is wrong. A steel door needs corrosion protection.
Answer C is wrong. A spliced frame rail needs corrosion protection.
Answer D is wrong. A spliced quarter panel needs corrosion protection.

Question #43
Answer A is wrong. Estimator A is right.
Answer B is wrong. Estimator B is right.
Answer C is correct.
Answer D is wrong.

Question #44
Answer A is wrong. Estimator A is right.
Answer B is wrong. Estimator B is right.
Answer C is correct.
Answer D is wrong.

Question #45
Answer A is wrong. Estimator A is right.
Answer B is correct. Estimator B is right.
Answer C is wrong.
Answer D is wrong.

Question #46
Answer A is correct. R&R stands for remove and replace.
Answer B is wrong. R&R stands for remove and replace.
Answer C is wrong. R&R stands for remove and replace.
Answer D is wrong. R&R stands for remove and replace.

Question #47
Answer A is wrong. Air bags can not be repacked.
Answer B is wrong. Not all air bag sensors need to be replaced. Check manufacturer's recommendations.
Answer C is wrong.
Answer D is correct.

Question #48
Answer A is wrong. The left fender is indicated.
Answer B is correct. The left fender is indicated.
Answer C is wrong. The left fender is indicated.
Answer D is wrong. The left fender is indicated.

Question #49
Answer A is wrong. A heating system contains a core.
Answer B is wrong. A heating system contains hoses.
Answer C is correct. A heating system does not contain an evaporator.
Answer D is wrong. A heating system contains controls.

Question #50
Answer A is wrong. Damaged high strength steel can not be repaired. It must be replaced.
Answer B is wrong. Kinked high strength steel can not be repaired. It must be replaced.
Answer C is wrong. Because A and B are wrong, then C is wrong.
Answer D is correct. Because A and B are wrong, then D is right.

Question #51
Answer A is correct. Decals are listed as a separate item.
Answer B is wrong. Solvent is additional material.
Answer C is wrong. Fasteners are additional material.
Answer D is wrong. Rust proofing is additional material.

Question #52
Answer A is wrong. Paint may be single stage.
Answer B is wrong. Paint may be basecoat/clearcoat.
Answer C is wrong. Paint may be tri-coat.
Answer D is correct. Paint may not be basecoat only.

Question #53
Answer A is wrong. The customer is the best source of information about the accident.
Answer B is wrong. The customer is the best source of information about the accident.
Answer C is wrong.
Answer D is correct.

Question #54
Answer A is wrong. The point of impact is direct damage, not indirect
Answer B is correct. The point of impact is direct damage.
Answer C is wrong.
Answer D is wrong.

Question #55
Answer A is wrong. Full frame is a type of vehicle construction.
Answer B is wrong. Unibody is a type of vehicle construction.
Answer C is correct. Unit frame is not a type of vehicle construction.
Answer D is wrong. Space frame is a type of vehicle construction.

Question #56
Answer A is wrong. A co-operative attitude will make the shop look professional.
Answer B is wrong. A co-operative attitude will develop a trusting relationship.
Answer C is correct. A co-operative attitude will not interfere with insurance company policies.
Answer D is wrong. A co-operative attitude will improve the customer's satisfaction.

Question #57
Answer A is correct. Estimator A is right.
Answer B is wrong. Estimator B is wrong. Commonly damaged engine parts are listed in a guide.
Answer C is wrong.
Answer D is wrong.

Question #58
Answer A is wrong. Estimator A is right.
Answer B is wrong. Estimator B is right.
Answer C is correct.
Answer D is wrong.

Question #59
Answer A is wrong. Discontinued parts are not available. Back-ordered parts may be available at some future time.
Answer B is correct. Discontinued parts are not available.
Answer C is wrong.
Answer D is wrong..

Question #60
Answer A is wrong. A kinked structural member must be replaced.
Answer B is correct. A kinked structural member must be replaced.
Answer C is wrong.
Answer D is wrong.

Question #61
Answer A is correct. The seat belts should be checked after a crash.
Answer B is wrong. The seat belts should be checked after a crash.
Answer C is wrong.
Answer D is wrong.

Question #62
Answer A is wrong. The arrow indicates the center link.
Answer B is wrong. The arrow indicates the center link.
Answer C is wrong. The arrow indicates the center link.
Answer D is correct. The arrow indicates the center link.

Question #63
Answer A is wrong. Estimator A is right.
Answer B is wrong. Estimator B is right.
Answer C is correct.
Answer D is wrong.

Question #64
Answer A is wrong. Add-on accessories are non-standard items and not listed in collision estimating guides.
Answer B is correct. Add-on accessories are non-standard items and not listed in collision estimating guides.
Answer C is wrong.
Answer D is wrong.

Question #65
Answer A is wrong. Estimator A is right.
Answer B is wrong. Estimator B is right.
Answer C is correct.
Answer D is wrong.

Question #66
Answer A is wrong. The customer should be given the warranty for sublet repairs and retain a copy for its records.
Answer B is correct. The customer should receive a written warranty.
Answer C is wrong.
Answer D is wrong.

Question #67
Answer A is wrong. Rust is only one type of pre-existing damage.
Answer B is wrong. Door dings are only one type of pre-existing damage.
Answer C is correct. Both rust and door dings are considered pre-existing damage.
Answer D is wrong.

Question #68
Answer A is correct. A shop should make all repairs on an estimate.
Answer B is wrong. Although many shops offer warranties, warranties are not required.
Answer C is wrong.
Answer D is wrong.

Question #69
Answer A is wrong. The minimum speed is 12 to 30 m.p.h., depending on the manufacturer.
Answer B is wrong. A continually glowing ABS light indicates a problem.
Answer C is wrong.
Answer D is correct.

Question #70
Answer A is wrong. Non-OEM parts are called aftermarket parts.
Answer B is correct. Non-OEM parts are called aftermarket parts.
Answer C is wrong. Non-OEM parts are called aftermarket parts.
Answer D is correct. Non-OEM parts are called aftermarket parts.

Question #71
Answer A is wrong. Estimator A is right.
Answer B is wrong. Estimator B is right.
Answer C is correct.
Answer D is wrong.

Question #72
Answer A is correct. A slide hammer is not used to repair frames.
Answer B is wrong. A floor pot system can be used to repair frames.
Answer C is wrong. A bench can be used to repair frames.
Answer D is wrong. A rack can be used to repair frames.

Question #73
Answer A is wrong. The condenser changes a gas to a liquid.
Answer B is wrong. The condenser changes a gas to a liquid.
Answer C is correct. The condenser changes a gas to a liquid.
Answer D is wrong. The condenser changes a gas to a liquid.

Question #74
Answer A is wrong. Estimator A is right.
Answer B is wrong. Estimator B is right.
Answer C is correct.
Answer D is wrong.

Question #75
Answer A is wrong. Estimator A is right.
Answer B is wrong. Estimator B is right.
Answer C is correct.
Answer D is wrong.

Question #76
Answer A is wrong. Estimator A is right.
Answer B is wrong. Estimator B is right.
Answer C is correct.
Answer D is wrong.

Question #77
Answer A is correct. The estimator and adjustor should review the estimate.
Answer B is wrong. The insurance estimate is not always the most appropriate way to repair the vehicle.
Answer C is wrong.
Answer D is wrong.

Question #78
Answer A is wrong. Estimator A is right.
Answer B is wrong. Estimator B is right.
Answer C is correct.
Answer D is wrong.

Question #79
Answer A is wrong. Estimator A is right.
Answer B is wrong. Estimator B is right.
Answer C is correct.
Answer D is wrong.

Question #80
Answer A is wrong. Estimator A is right.
Answer B is wrong. Estimator B is right.
Answer C is correct.
Answer D is wrong.

Question #81
Answer A is wrong. A bumper shock should be replaced if it has a leaking tube.
Answer B is wrong. A bumper shock should be replaced if it has a bent tube.
Answer C is correct. A minor bend in the mounting plate can be repaired.
Answer D is wrong. A bumper shock should be replaced if it has a broken tube.

Question #82
Answer A is correct. The alternator converts mechanical energy into electrical energy.
Answer B is wrong. The alternator converts mechanical energy into electrical energy.
Answer C is wrong.
Answer D is wrong.

Question #83
Answer A is correct. The control unit might not be the problem but the entire system should be checked by the collision repair shop.
Answer B is wrong. The control unit might not be the problem but the entire system should be checked by the collision repair shop.
Answer C is wrong.
Answer D is wrong.

Question #84
Answer A is wrong. 4.0 hr + 2.5 hr + 1.0 hr = 7.5 hrs.
Answer B is wrong. 4.0 hr + 2.5 hr + 1.0 hr = 7.5 hrs.
Answer C is correct. 4.0 hr + 2.5 hr + 1.0 hr = 7.5 hrs.
Answer D is wrong. 4.0 hr + 2.5 hr + 1.0 hr = 7.5 hrs.

Question #85
Answer A is wrong. The "P" pages do not list refinish time.
Answer B is correct. The head notes list refinish time.
Answer C is wrong.
Answer D is wrong.

Question #86
Answer A is wrong. Estimator A is right.
Answer B is wrong. Estimator B is right.
Answer C is correct.
Answer D is wrong.

Question #87
Answer A is wrong. Estimator A is right.
Answer B is wrong. Estimator B is right.
Answer C is correct.
Answer D is wrong.

Question #88
Answer A is correct. Although a unibody vehicle may have diamond, the damage would be so extensive it would not be repairable.
Answer B is wrong. A unibody may have twist.
Answer C is wrong. A unibody may have sag.
Answer D is wrong. A unibody may have mash.

Question #89
Answer A is wrong. Used parts do not cost as much as OEM parts.
Answer B is correct. Used parts are usually sold as assemblies.
Answer C is wrong.
Answer D is wrong.

Question #90
Answer A is wrong. Estimator A is right.
Answer B is wrong. Estimator B is right.
Answer C is correct.
Answer D is wrong.

Question #91
Answer A is wrong. Estimator A is right.
Answer B is wrong. Estimator B is right.
Answer C is correct.
Answer D is wrong.

Question #92
Answer A is wrong. A rear clip is a salvage assembly.
Answer B is correct. A bumper cover is a single part, not an assembly.
Answer C is wrong. A rear clip top is a salvage assembly.
Answer D is wrong. A front clip is a salvage assembly.

Question #93
Answer A is wrong. A tram gauge can not be used to measure the center line.
Answer B is wrong. A tram gauge can not be used to measure height.
Answer C is wrong.
Answer D is correct.

Question #94
Answer A is correct. A non-functioning sensor at the indicated location could cause a no-start.
Answer B is wrong. A knock sensor does not cause a no-start.
Answer C is wrong.
Answer D is wrong.

Question #95
Answer A is correct. Integrity is considered when evaluating a repair.
Answer B is wrong. The cost is not considered as part of quality.
Answer C is wrong. The speed of the repair is not considered as part of quality.
Answer D is wrong. The location is not considered as part of quality.

Question #96
Answer A is wrong. Estimator A is right.
Answer B is wrong. Estimator B is right.
Answer C is correct.
Answer D is wrong.

Question #97
Answer A is correct. Not all listed aftermarket parts are available.
Answer B is wrong. An OEM dealer would have no information about aftermarket parts.
Answer C is wrong.
Answer D is wrong.

Question #98
Answer A is wrong. Estimator A is right.
Answer B is wrong. Estimator B is right.
Answer C is correct.
Answer D is wrong.

Question #99
Answer A is wrong. Slight damage does not exceed cost of door shell replacement.
Answer B is wrong. Slight damage does not exceed cost of door skin replacement.
Answer C is wrong.
Answer D is correct.

Question #100
Answer A is wrong. Only if the brake shoes are worn, should they be replaced.
Answer B is correct. Because the brake line must be disconnected to replace the backing plate, the brakes must be bled. An allowance is needed.
Answer C is wrong.
Answer D is wrong.

Question #101
Answer A is wrong.
Answer B is wrong.
Answer C is correct. VIN means Vehicle Identification Number.
Answer D is wrong.

Question #102
Answer A is wrong. Estimator A is right.
Answer B is wrong. Estimator B is right.
Answer C is correct.
Answer D is wrong.

Question #103
Answer A is wrong.
Answer B is wrong. .
Answer C is wrong. .
Answer D is correct. Metric bolts have numbers indicating strength.

Question #104
Answer A is wrong. A disconnected vacuum line causes a rough idle on older vehicles.
Answer B is wrong. A disconnected vacuum line causes a high idle on newer vehicles.
Answer C is wrong.
Answer D is correct.

Question #105
Answer A is wrong.
Answer B is correct. The part pictured is a bumper cover.
Answer C is wrong.
Answer D is wrong.

Answers to the Test Questions for the Additional Test Questions Section 6

1. D
2. B
3. A
4. C
5. C
6. B
7. C
8 B
9. C
10. A
11. A
12. B
13. C
14. C
15. C
16. A
17. C
18. D
19. B
20. C
21. B
22. C
23. A
24. A
25. C
26. C
27. D
28. B
29. D
30. C
31. B
32. C
33. C
34. C
35. A
36. C
37. A
38. D
39. D
40. B
41. D
42. A
43. D
44. C
45. C
46. B
47. C
48. C
49. A
50. C
51. D
52. A
53. C
54. C
55. D
56. A
57. D
58. C
59. A
60. C
61. D
62. B
63. B
64. C
65. A
66. C
67. B
68. B
69. D
70. C
71. D
72. D
73. B
74. C
75. A
76. A
77. C
78. D
79. C
80. A
81. A
82. A
83. C
84. B
85. C
86. C
87. C
88. C
89. C
90. C
91. A
92. B
93. A
94. A
95. D
96. C
97. C
98. C
99. D
100. B
101. A
102. C
103. A
104. A
105. D
106. A
107. D
108. C
109. A
110. C
111. C
112. B
113. C
114. C
115. C
116. C
117. C
118. C
119. A
120. D
121. C
122. C
123. B
124. B
125. C
126. D
127. D
128. C
129. C
130. C
131. B
132. D
133. C
134. B
135. C
136. B
137. B
138. A
139. B
140. B
141. A
142. A
143. B
144. A
145. D
146. B
147. B
148. D
149. A
150. A
151. C
152. A

153. D
154. B
155. C
156. A
157. C
158. B
159. D
160. A
161. C
162. B
163. C
164. C
165. B
166. B
167. D
168. B
169. A
170. A
171. B
172. C
173. C
174. A
175. B
176. D
177. C
178. C
179. C
180. C
181. B
182. B
183. A
184. B
185. C
186. C
187. C
188. A
189. C
190. B
191. D
192. C
193. B
194. C
195. C
196. B
197. C
198. C
199. C
200. C
201. B
202. A
203. C
204. C
205. A
206. B
207. A
208. C
209. D
210. A

Explanations to the Answers for the Additional Test Questions Section 6

Question #1
Answer A is wrong. Full attention will make the shop look professional.
Answer B is wrong. Full attention will develop a trusting relationship.
Answer C is wrong. Full attention will improve customer satisfaction.
Answer D is correct. Full attention will not irritate other customers.

Question #2
Answer A is wrong. Replacement fenders need to painted.
Answer B is correct. Bumper reinforcements are not refinished.
Answer C is wrong. Replacement fenders need to be painted.
Answer D is wrong. Replacement lower frame rails need to be painted.

Question #3
Answer A is correct. Included operations are part of another labor operation.
Answer B is wrong. Alignment other than adjustment is not part of overhaul.
Answer C is wrong.
Answer D is wrong.

Question #4
Answer A is wrong. A is right.
Answer B is wrong. B is right.
Answer C is correct.
Answer D is wrong.

Question #5
Answer A is wrong. A lift may be needed to see the underbody.
Answer B is wrong. A trouble light may be needed to illuminate the dark areas.
Answer C is correct. A computer estimate system is not needed to inspect a damaged vehicle.
Answer D is wrong. A clean stall is much easier to work in.

Question #6
Answer A is wrong. A parallelogram steering system may include a tie rod.
Answer B is correct. A parallelogram steering system does not include a rack and pinion.
Answer C is wrong. A parallelogram steering system may include a center link.
Answer D is wrong. A parallelogram steering system may include a drag link.

Question #7
Answer A is wrong. A is right.
Answer B is wrong. B is right.
Answer C is correct.
Answer D is wrong.

Question #8
Answer A is wrong. Always thoroughly explain the repair process to the customer.
Answer B is correct. Always thoroughly explain the repair process to the customer.
Answer C is wrong.
Answer D is wrong.

Question #9
Answer A is wrong. A is right.
Answer B is wrong. B is right.
Answer C is correct.
Answer D is wrong.

Question #10
Answer A is correct. Customers should always be greeted in a friendly manner.
Answer B is wrong. Customers should always be greeted in a friendly manner.
Answer C is wrong.
Answer D is wrong.

Question #11
Answer A is correct. Manual door locks are not part of the electrical system.
Answer B is wrong. Power door locks are part of the electrical system.
Answer C is wrong. Batteries are part of the electrical system.
Answer D is wrong. Head lights are part of the electrical system.

Question #12
Answer A is wrong. The insurance company does not pay for betterment.
Answer B is correct. The customer pays for betterment.
Answer C is wrong.
Answer D is wrong.

Question #13
Answer A is wrong. A is right.
Answer B is wrong. B is right.
Answer C is correct.
Answer D is wrong.

Question #14
Answer A is wrong. A is right.
Answer B is wrong. B is right.
Answer C is correct.
Answer D is wrong.

Question #15
Answer A is wrong. A is right.
Answer B is wrong. B is right.
Answer C is correct.
Answer D is wrong.

Question #16
Answer A is correct. A vapor canister is not located in the glove box.
Answer B is wrong. A vapor canister may be near the gas tank.
Answer C is wrong. A vapor canister may be on the apron.
Answer D is wrong. A vapor canister may be in the engine compartment.

Question #17
Answer A is wrong. R&I means that the windshield will be taken out, then put back in place or installed.
Answer B is wrong. R&I means that the windshield will be taken out, then put back in place.
Answer C is correct. R&I means that the windshield will be taken out, then put back in place.
Answer D is wrong. R&I means that the windshield will be taken out, then put back in place.

Question #18
Answer A is wrong. The letters R&I stand for remove and install.
Answer B is wrong. The letters R&I stand for remove and install.
Answer C is wrong. The letters R&I stand for remove and install.
Answer D is correct. The letters R&I stand for remove and install.

Question #19
Answer A is wrong. A/C recharge consists of adding oil and refrigerant.
Answer B is correct. A/C recharge consists of adding oil and refrigerant.
Answer C is wrong. A/C recharge consists of adding oil and refrigerant.
Answer D is wrong. A/C recharge consists of adding oil and refrigerant.

Question #20
Answer A is wrong. A is right.
Answer B is wrong. B is right.
Answer C is correct.
Answer D is wrong.

Question #21
Answer A is wrong. A re-manufactured bumper costs less than OEM.
Answer B is correct. A re-manufactured bumper costs less than OEM.
Answer C is wrong.
Answer D is wrong.

Question #22
Answer A is wrong. Estimator A is correct.
Answer B is wrong. Estimator B is correct.
Answer C is correct.
Answer D is wrong.

Question #23
Answer A is correct. Not all aftermarket parts listed are available.
Answer B is wrong. Not all aftermarket parts listed are available.
Answer C is wrong.
Answer D is wrong.

Question #24
Answer A is correct. Lower frame rails are usually not available as aftermarket.
Answer B is wrong. Fenders are usually available as aftermarket.
Answer C is wrong. Bumpers are usually available as aftermarket.
Answer D is wrong. Hoods are usually available as aftermarket.

Question #25
Answer A is wrong. A is right.
Answer B is wrong. B is right.
Answer C is correct.
Answer D is wrong.

Question #26
Answer A is wrong. Estimator A is correct.
Answer B is wrong. Estimator B is correct.
Answer C is correct.
Answer D is wrong.

Question #27
Answer A is wrong. Bumper removal is not included.
Answer B is wrong. Hole drilling is not included.
Answer C is wrong. Refinishing is not included.
Answer D is correct. Side marker lamp replacement is included.

Question #28
Answer A is wrong. The part is the lower frame rail.
Answer B is correct. The part is the lower frame rail.
Answer C is wrong. The part is the lower frame rail.
Answer D is wrong. The part is the lower frame rail.

Question #29
Answer A is wrong. A front wheel drive's drive train includes a transaxle.
Answer B is wrong. A front wheel drive's drive train includes an engine.
Answer C is wrong. A front wheel drive's drive train includes half shafts.
Answer D is correct. A front wheel drive's drive train does not include a transmission.

Question #30
Answer A is wrong. The assemblies are listed front to back.
Answer B is wrong. The assemblies are listed front to back.
Answer C is correct. The assemblies are listed front to back.
Answer D is wrong. The assemblies are listed front to back.

Question #31
Answer A is wrong. In space frame construction, the outer body panels do not contribute to overall strength.
Answer B is correct. In space frame construction, the outer body panels do not contribute to overall strength.
Answer C is wrong.
Answer D is wrong.

Question #32
Answer A is wrong. Estimator A is correct.
Answer B is wrong. Estimator B is correct.
Answer C is correct.
Answer D is wrong.

Question #33
Answer A is wrong. A is right.
Answer B is wrong. B is right.
Answer C is correct.
Answer D is wrong.

Question #34
Answer A is wrong. A is right.
Answer B is wrong. B is right.
Answer C is correct.
Answer D is wrong.

Question #35
Answer A is correct. The air bag light will glow if the bag is deployed.
Answer B is wrong. The air bag light will glow if the bag is deployed.
Answer C is wrong.
Answer D is wrong.

Question #36
Answer A is wrong. A plastic urethane bumper does not need corrosion protection.
Answer B is wrong. A plastic SMC door does not need corrosion protection.
Answer C is correct. A steel fender needs corrosion protection.
Answer D is wrong. A plastic RRIM fender does not need corrosion protection.

Question #37
Answer A is correct. OSHA requires that workers be informed of hazards.
Answer B is wrong. OSHA, not EPA, regulates the workers' health.
Answer C is wrong.
Answer D is wrong.

Question #38
Answer A is wrong. Upgrade is not the term.
Answer B is wrong. Depreciate is not the term.
Answer C is wrong. Stabilize is not the term.
Answer D is correct. Betterment describes improving a damaged vehicle to better than pre-accident condition.

Question #39
Answer A is wrong. No damaged steering components should be straightened.
Answer B is wrong. No damaged suspension components should be straightened.
Answer C is wrong.
Answer D is correct.

Question #40
Answer A is wrong. Repairs to safety systems are the most important of the four.
Answer B is correct. Repairs to safety systems are the most important of the four.
Answer C is wrong. Repairs to safety systems are the most important of the four.
Answer D is wrong. Repairs to safety systems are the most important of the four.

Question #41
Answer A is wrong. Crash sensors may not need to be replaced.
Answer B is wrong. The clock spring may not need to be replaced.
Answer C is wrong. The dash panel may not need to be replaced.
Answer D is correct. The air bag module must always be replaced.

Question #42
Answer A is correct. Frame measurement time includes the time to diagnose.
Answer B is wrong. Frame pulling is not included in frame measurement time.
Answer C is wrong.
Answer D is wrong.

Question #43
Answer A is wrong. A strut is part of the knee.
Answer B is wrong. A knuckle is part of the knee.
Answer C is wrong. A control arm is part of the knee.
Answer D is correct. A half shaft is not part of the knee.

Question #44
Answer A is wrong. The automatic seat belt system places the shoulder belt on the occupant.
Answer B is wrong. The automatic seat belt system places the shoulder belt on the occupant.
Answer C is correct. The automatic seat belt system places the shoulder belt on the occupant.
Answer D is wrong. The automatic seat belt system places the shoulder belt on the occupant.

Question #45
Answer A is wrong. A is right.
Answer B is wrong. B is right.
Answer C is correct.
Answer D is wrong.

Question #46
Answer A is wrong. An arc welder can not be used to weld aluminum.
Answer B is correct. A TIG welder can be used to weld aluminum.
Answer C is wrong.
Answer D is wrong.

Question #47
Answer A is wrong. A is right.
Answer B is wrong. B is right.
Answer C is correct.
Answer D is wrong.

Question #48
Answer A is wrong. The arrow indicates the pitman arm.
Answer B is wrong. The arrow indicates the pitman arm.
Answer C is correct. The arrow indicates the pitman arm.
Answer D is wrong. The arrow indicates the pitman arm.

Question #49
Answer A is correct. A properly sectioned vehicle is crash worthy.
Answer B is wrong. A properly sectioned vehicle is durable.
Answer C is wrong.
Answer D is wrong.

Question #50
Answer A is wrong. A is right.
Answer B is wrong. B is right.
Answer C is correct.
Answer D is wrong.

Question #51
Answer A is wrong. Standard markup is 25%.
Answer B is wrong. Standard markup is 25%.
Answer C is wrong.
Answer D is correct.

Question #52
Answer A is correct. A brake warning light glows if the brake system loses pressure.
Answer B is wrong. The ABS warning light, not the brake warning light, glows if the ABS system has a malfunction.
Answer C is wrong.
Answer D is wrong.

Question #53
Answer A is wrong. Overhaul includes part removal.
Answer B is wrong. Overhaul includes part installation.
Answer C is correct. Overhaul does not include refinish.
Answer D is wrong. Overhaul includes inspection.

Question #54
Answer A is wrong. The availability of a back-ordered part is indeterminate, could be a week or a month.
Answer B is wrong. The availability of a back-ordered part is indeterminate.
Answer C is correct.
Answer D is wrong.

Question #55
Answer A is wrong. A vacuum leak on a newer car may cause a high idle.
Answer B is wrong. Only if recommended by the manufacturer should additives be used.
Answer C is wrong. Because A and B are wrong, then C is wrong.
Answer D is correct. Because A and B are wrong, then D is correct.

Question #56
Answer A is correct. A is right.
Answer B is wrong. B is right.
Answer C is wrong.
Answer D is wrong.

Question #57
Answer A is wrong. Calling the customer can be used to answer questions.
Answer B is wrong. Calling the customer can be used to ask about pre-existing damage.
Answer C is wrong. Calling the customer can be used to explain the repair.
Answer D is correct. Calling the customer should not be used to complain about the insurance company.

Question #58
Answer A is wrong. A is right.
Answer B is wrong. B is right.
Answer C is correct.
Answer D is wrong.

Question #59
Answer A is correct. The same type of fastener should be used as a replacement.
Answer B is wrong. The same type of fastener should be used as a replacement.
Answer C is wrong.
Answer D is wrong.

Question #60
Answer A is wrong. A is right.
Answer B is wrong. B is right.
Answer C is correct.
Answer D is wrong.

Question #61
Answer A is wrong. A broken styrofoam impact absorber should be replaced, not repaired.
Answer B is wrong. A dented styrofoam impact absorber should be replaced, not repaired.
Answer C is wrong.
Answer D is correct.

Question #62
Answer A is wrong. The picture is a unibody.
Answer B is correct. The picture is a unibody.
Answer C is wrong. The picture is a unibody.
Answer D is wrong. The picture is a unibody.

Question #63
Answer A is wrong. Customers are usually interested in the repair.
Answer B is correct. Calling the customer keeps him involved.
Answer C is wrong.
Answer D is wrong.

Question #64
Answer A is wrong. Estimator A is correct.
Answer B is wrong. Estimator B is correct.
Answer C is correct.
Answer D is wrong.

Question #65
Answer A is correct. The estimator must listen to the customer.
Answer B is wrong. The customer may be able to point out damage of which the estimator is unaware.
Answer C is wrong.
Answer D is wrong.

Question #66
Answer A is wrong. A is right.
Answer B is wrong. B is right.
Answer C is correct.
Answer D is wrong.

Question #67
Answer A is wrong. If the brake shoes are not worn, they do not need to be replaced.
Answer B is correct. If the brake shoes are not worn, they do not need to be replaced.
Answer C is wrong.
Answer D is wrong.

Question #68
Answer A is wrong. Production date should be recorded.
Answer B is correct. Gas mileage is not relevant.
Answer C is wrong. Mileage should be recorded.
Answer D is wrong. License plate number should be recorded.

Question #69
Answer A is wrong. Safety related items are the most commonly used OEM parts.
Answer B is wrong. Safety related items are the most commonly used OEM parts.
Answer C is wrong. Safety related items are the most commonly used OEM parts.
Answer D is correct. Safety related items are the most commonly used OEM parts.

Question #70
Answer A is wrong. A is right.
Answer B is wrong. B is right.
Answer C is correct.
Answer D is wrong.

Question #71
Answer A is wrong. Air bags are safety-related.
Answer B is wrong. ABS is safety-related.
Answer C is wrong. Tie rod is safety-related.
Answer D is correct. A fender is not safety-related.

Question #72
Answer A is wrong. The letters VOC stand for volatile organic compound.
Answer B is wrong. The letters VOC stand for volatile organic compound.
Answer C is wrong. The letters VOC stand for volatile organic compound.
Answer D is correct. The letters VOC stand for volatile organic compound.

Question #73
Answer A is wrong. R-12 and R-134a can not be interchanged.
Answer B is correct. R-12 and R-134a can not be interchanged.
Answer C is wrong.
Answer D is wrong.

Question #74
Answer A is wrong. A is right.
Answer B is wrong. B is right.
Answer C is correct.
Answer D is wrong.

Question #75
Answer A is correct. A deployed air bag must be replaced.
Answer B is wrong. Depending on the manufacturer's recommendations, sensors may not need replacement after deployment.
Answer C is wrong.
Answer D is wrong.

Question #76
Answer A is correct. Rear doors are not part of a rear clip.
Answer B is wrong. The deck lid is part of a rear clip.
Answer C is wrong. The tail lights are part of a rear clip.
Answer D is wrong. The rear bumper is part of a rear clip.

Question #77
Answer A is wrong. Percentage of the new price is not an accurate way to determine the used car price.
Answer B is wrong. The owner's manual does not contain any information on pricing
Answer C is correct. A used car value guide determines the vehicle's price.
Answer D is wrong. A collision estimating guide does not contain vehicle price information.

Question #78
Answer A is wrong. Both unibody and full frames can have mash damage.
Answer B is wrong. Both unibody and full frames can have mash damage.
Answer C is wrong.
Answer D is correct.

Question #79
Answer A is wrong. Body filler is a body repair material.
Answer B is wrong. MIG weld wire is a body repair material.
Answer C is correct. Clearcoat is a refinishing material.
Answer D is wrong. Plastic repair material is a body repair material.

Question #80
Answer A is correct. Sublet repairs are warranted by the sublet shop.
Answer B is wrong. Parts are warranted by the shop.
Answer C is wrong. Paint is warranted by the shop.
Answer D is wrong. Workmanship is warranted by the shop.

Question #81
Answer A is correct. Brake bleeding is required whenever a brake system is opened.
Answer B is wrong. Brake bleeding is required whenever a brake system is opened.
Answer C is wrong.
Answer D is wrong.

Question #82
Answer A is correct. Estimator A is right, record the customer information first.
Answer B is wrong. Even though the paint code is useful, record the customer information first.
Answer C is wrong.
Answer D is wrong.

Question #83
Answer A is wrong. A is right.
Answer B is wrong. B is right.
Answer C is correct.
Answer D is wrong.

Question #84
Answer A is wrong. If frame damage is not suspected, measuring time is not needed.
Answer B is correct. Vehicles with suspected frame damage, should have measuring time.
Answer C is wrong.
Answer D is wrong.

Question #85
Answer A is wrong. Steering gear box damage is not indicated by jouncing.
Answer B is wrong. Jouncing is only one part of front end alignment procedure.
Answer C is correct. Steering rack alignment is tested by jouncing.
Answer D is wrong. Steering arm damage is not indicated by jouncing.

Question #86
Answer A is wrong. The figure shows an A/C system.
Answer B is wrong. The figure shows an A/C system.
Answer C is correct. The figure shows an A/C system.
Answer D is wrong. The figure shows an A/C system.

Question #87
Answer A is wrong. A is right.
Answer B is wrong. B is right.
Answer C is correct.
Answer D is wrong.

Question #88
Answer A is wrong. A is right.
Answer B is wrong. B is right.
Answer C is correct.
Answer D is wrong.

Question #89
Answer A is wrong. A is right.
Answer B is wrong. B is right.
Answer C is correct.
Answer D is wrong.

Question #90
Answer A is wrong. This statement is correct.
Answer B is wrong. This statement is also correct.
Answer C is correct.
Answer D is wrong.

Question #91
Answer A is correct. A positive attitude is important in winning trust.
Answer B is wrong. A messy desk gives the impression of disorganization.
Answer C is wrong.
Answer D is wrong.

Question #92
Answer A is wrong. Each employee that answers the shop phone should identify the shop and himself.
Answer B is correct. Each employee that answers the shop phone should identify the shop and himself.
Answer C is wrong.
Answer D is wrong.

Question #93
Answer A is correct. Always check for hidden damage, even on a light hit.
Answer B is wrong. Even a light front hit can have hidden damage.
Answer C is wrong.
Answer D is wrong.

Question #94
Answer A is correct. Unloading the suspension is not part of centering the steering wheel.
Answer B is wrong. Turning the steering wheel lock to lock is part of centering the steering wheel.
Answer C is wrong. Counting the number of turns is part of centering the steering wheel.
Answer D is wrong. Dividing the number of turns by 2 is part of centering the steering wheel.

Question #95
Answer A is wrong. The usual markup is 25%.
Answer B is wrong. The usual markup is 25%.
Answer C is wrong. The usual markup is 25%.
Answer D is correct. The usual markup is 25%.

Question #96
Answer A is wrong. First panel, no deduction, each other adjacent panel - 0.4 hr. per panel - 0.4 x 3 = 1.2
Answer B is wrong. First panel, no deduction, each other adjacent panel - 0.4 hr. per panel.
Answer C is correct. First panel, no deduction, each other adjacent panel - 0.4 hr. per panel.
Answer D is wrong. First panel, no deduction, each other adjacent panel - 0.4 hr. per panel.

Question #97
Answer A is wrong. A is right.
Answer B is wrong. B is right.
Answer C is correct.
Answer D is wrong.

Question #98
Answer A is wrong. A is right.
Answer B is wrong. B is right.
Answer C is correct.
Answer D is wrong.

Question #99
Answer A is wrong. The customer is usually aware of the insurance company.
Answer B is wrong. The customer usually knows his work phone number.
Answer C is wrong. The customer usually knows his home address.
Answer D is correct. The customer may not be aware of the claim number and if the repairs are not paid by insurance, there will be no claim number.

Question #100
Answer A is wrong. A re-manufactured bumper is also called a rechromed bumper.
Answer B is correct. A re-manufactured bumper is also called a rechromed bumper.
Answer C is wrong. A re-manufactured bumper is also called a rechromed bumper.
Answer D is wrong. A re-manufactured bumper is also called a rechromed bumper.

Question #101
Answer A is correct. "O" rings should be replaced when A/C lines are disconnected.
Answer B is wrong. Refrigerant must be recovered, not released.
Answer C is wrong. Because B is wrong, then C is wrong.
Answer D is wrong. Because A is right, then D is wrong.

Question #102
Answer A is wrong. Estimator A is right.
Answer B is wrong. Estimator B is right.
Answer C is correct.
Answer D is wrong.

Question #103
Answer A is correct. An accumulator is also called a drier.
Answer B is wrong. An evaporator is not called a drier.
Answer C is wrong.
Answer D is wrong.

Question #104
Answer A is correct. Always check for hidden damage. Removing the fender will help see cowl damage.
Answer B is wrong. Even a light front hit can have hidden damage.
Answer C is wrong.
Answer D is wrong.

Question #105
Answer A is wrong. The fender is the most likely of these parts to be available as aftermarket.
Answer B is wrong. The fender is the most likely of these parts to be available as aftermarket.
Answer C is wrong. The fender is the most likely of these parts to be available as aftermarket.
Answer D is correct. The fender is the most likely of these parts to be available as aftermarket.

Question #106
Answer A is correct. A kink is a bend of greater than 90 degrees over a short radius
Answer B is wrong. A bend of less than 90 degrees is not a kink
Answer C is wrong. A kink is over a short radius
Answer D is wrong. A kink is greater than 90 degrees

Question #107
Answer A is wrong. The shop determines how long the warranty is. There is no specified duration for all warranties.
Answer B is wrong. The shop determines how long the warranty is.
Answer C is wrong.
Answer D is correct.

Question #108
Answer A is wrong. A is right.
Answer B is wrong. B is right.
Answer C is correct.
Answer D is wrong.

Question #109
Answer A is correct. Add-on accessory labor time is not included in collision estimating guides.
Answer B is wrong. OEM part labor times is included in collision estimating guides.
Answer C is wrong. OEM part refinish time is included in collision estimating guides.
Answer D is wrong. OEM part overhaul time is included in collision estimating guides.

Question #110
Answer A is wrong. A is right.
Answer B is wrong. B is right.
Answer C is correct.
Answer D is wrong.

Question #111
Answer A is wrong. A is right.
Answer B is wrong. B is right.
Answer C is correct.
Answer D is wrong.

Question #112
Answer A is wrong. Even though initially options may not seem to be needed, they may be needed to order parts.
Answer B is correct. The presence of safety systems should be noted in the estimate.
Answer C is wrong.
Answer D is wrong.

Question #113
Answer A is wrong. A is right.
Answer B is wrong. B is right.
Answer C is correct.
Answer D is wrong.

Question #114
Answer A is wrong. A is right.
Answer B is wrong. B is right.
Answer C is correct.
Answer D is wrong.

Question #115
Answer A is wrong.
Answer B is wrong.
Answer C is correct. Lines or dots are used to indicate the strength on a standard bolt.
Answer D is wrong.

Question #116
Answer A is wrong. Objects in the trunk may damage the deck lid in a roll over.
Answer B is wrong. Objects in the trunk may damage the quarter panel in a side impact.
Answer C is correct. Objects inside the vehicle do not damage the fender.
Answer D is wrong. Objects inside the vehicle may damage the door in a side impact.

Question #117
Answer A is wrong. A is right.
Answer B is wrong. B is right.
Answer C is correct.
Answer D is wrong.

Question #118
Answer A is wrong. A is right.
Answer B is wrong. B is right.
Answer C is correct.
Answer D is wrong.

Question #119
Answer A is correct. Ultra high strength steel is used to make door crash beams.
Answer B is wrong. Fenders are not made from ultra high strength steel.
Answer C is wrong. Hoods are not made from ultra high strength steel.
Answer D is wrong. Quarter panels are not made from ultra high strength steel.

Question #120
Answer A is wrong. The drawing is of a fuel injector.
Answer B is wrong. The drawing is of a fuel injector.
Answer C is wrong. The drawing is of a fuel injector.
Answer D is correct. The drawing is of a fuel injector.

Question #121
Answer A is wrong. Because the idler arm is not adjustable, an alignment is not required.
Answer B is wrong. Because a tie rod is adjustable, an alignment is required.
Answer C is correct.
Answer D is wrong.

Question #122
Answer A is wrong. The evaporator changes refrigerant gas to liquid.
Answer B is wrong. The evaporator changes refrigerant gas to liquid.
Answer C is correct. The evaporator changes refrigerant gas to liquid.
Answer D is wrong. The evaporator changes refrigerant gas to liquid.

Question #123
Answer A is wrong. A glowing ABS warning light indicates malfunction.
Answer B is correct. A glowing ABS warning light indicates malfunction.
Answer C is wrong.
Answer D is wrong.

Question #124
Answer A is wrong. The pitman arm connects the steering gear box to the center link.
Answer B is correct. The pitman arm connects the steering gear box to the center link.
Answer C is wrong. The pitman arm connects the steering gear box to the center link.
Answer D is wrong. The pitman arm connects the steering gear box to the center link.

Question #125
Answer A is wrong. A is right.
Answer B is wrong. B is right.
Answer C is correct.
Answer D is wrong.

Question #126
Answer A is wrong. Unibody vehicles do not have a separate frame.
Answer B is wrong. Space frame vehicles similar to unibody vehicles.
Answer C is wrong.
Answer D is correct.

Question #127
Answer A is wrong. Usually a bench takes longer to set up than a floor pot system.
Answer B is wrong. Benches may be used for heavy damage.
Answer C is wrong.
Answer D is correct.

Question #128
Answer A is wrong. A is right.
Answer B is wrong. B is right.
Answer C is correct.
Answer D is wrong.

Question #129
Answer A is wrong. Inspection is one source of information.
Answer B is wrong. Customer interview is one source of information.
Answer C is correct.
Answer D is wrong.

Question #130
Answer A is wrong. The indicated part is the center pillar.
Answer B is wrong. The indicated part is the center pillar.
Answer C is correct. The indicated part is the center pillar.
Answer D is wrong. The indicated part is the center pillar.

Question #131
Answer A is wrong. Damaged suspension components must be replaced.
Answer B is correct. The most cost effective repair method must be determined.
Answer C is wrong.
Answer D is wrong.

Question #132
Answer A is wrong. Fenders are commonly available as aftermarket.
Answer B is wrong. Bumpers are commonly available as aftermarket.
Answer C is wrong. Header panels are commonly available as aftermarket.
Answer D is correct. Frame rails are not usually available as aftermarket.

Question #133
Answer A is wrong. A two-channel ABS controls the rear wheels independently.
Answer B is wrong. A two-channel ABS controls the rear wheels independently.
Answer C is correct.
Answer D is wrong.

Question #134
Answer A is wrong. A urethane bumper cover will not corrode.
Answer B is correct. A welded area is a corrosion hot spot.
Answer C is wrong. A steel fender will corrode but not as fast as a weld.
Answer D is wrong. An SMC hatch will not corrode.

Question #135
Answer A is wrong. A is right.
Answer B is wrong. B is right.
Answer C is correct.
Answer D is wrong.

Question #136
Answer A is wrong. Aftermarket parts may need holes drilled.
Answer B is correct. Aftermarket parts may need holes drilled.
Answer C is wrong.
Answer D is wrong.

Question #137
Answer A is wrong. The head light is likely damaged in a frontal collision.
Answer B is correct. The starter is the least likely of the listed items to be damaged.
Answer C is wrong. The turn signal lamp is likely damaged in a frontal collision.
Answer D is wrong. The battery is more likely to be damaged than the starter.

Question #138
Answer A is correct. Confidence in the shop can be created by calling the customer.
Answer B is wrong. The customer, not the insurance company, will have information on pre-existing conditions.
Answer C is wrong.
Answer D is wrong.

Question #139
Answer A is wrong. The left quarter panel is indicated.
Answer B is correct. The left quarter panel is indicated.
Answer C is wrong. The left quarter panel is indicated.
Answer D is wrong. The left quarter panel is indicated.

Question #140
Answer A is wrong. The customer need not know the advantages of a computer-generated estimate.
Answer B is correct. A complete estimate gives the customer the best idea of repair cost.
Answer C is wrong.
Answer D is wrong.

Question #141
Answer A is correct. The customer should be informed when the vehicle will be repaired.
Answer B is wrong. The customer should be encouraged to call with questions.
Answer C is wrong.
Answer D is wrong.

Question #142
Answer A is correct. Overlap must be figured into refinish times.
Answer B is wrong. Included items should not be added to the estimate.
Answer C is wrong.
Answer D is wrong.

Question #143
Answer A is wrong. Stripe tape should be listed as a separate item.
Answer B is correct. Body filler should be included in additional material.
Answer C is wrong. Body side molding should be listed as a separate item.
Answer D is wrong. Decals should be listed as a separate item.

Question #144
Answer A is correct. 7.5 hr. x $20/hr = $150.00
Answer B is wrong. 7.5 hr. x $20/hr = $150.00
Answer C is wrong. 7.5 hr. x $20/hr = $150.00
Answer D is wrong. 7.5 hr. x $20/hr = $150.00

Question #145
Answer A is wrong. A survey is good business practice.
Answer B is wrong. A follow-up call is good business practice.
Answer C is wrong. Asking questions is good business practice.
Answer D is correct. The vehicle should be cleaned and ready before the customer arrives.

Question #146
Answer A is wrong. A laser can be used to measure center line
Answer B is correct. A tram gauge can not be used to measure center line
Answer C is wrong. A universal bench can be used to measure center line
Answer D is wrong. Self centering gauge can be used to measure center line

Question #147
Answer A is wrong. Body filler will not adhere to flexible plastic.
Answer B is correct. Adhesives can be used to repair flexible plastic.
Answer C is wrong.
Answer D is wrong.

Question #148
Answer A is wrong. Not identifying yourself on the phone does not create trust.
Answer B is wrong. Not identifying yourself on the phone does not instill confidence.
Answer C is wrong. Not identifying yourself on the phone is not professional.
Answer D is correct. Not identifying yourself on the phone does create a question of honesty.

Question #149
Answer A is correct. Standard bolts have lines or dots on the head.
Answer B is wrong. Metric bolts have numbers.
Answer C is wrong.
Answer D is wrong.

Question #150
Answer A is correct. Some sublet suppliers charge the body shop less than the standard fee.
Answer B is wrong. Bumper replacement is not usually sublet.
Answer C is wrong.
Answer D is wrong.

Question #151
Answer A is wrong. A is right.
Answer B is wrong. B is right.
Answer C is correct.
Answer D is wrong.

Question #152
Answer A is correct. Starting at the point of impact is the best way to evaluate damage.
Answer B is wrong. The point furthest from the damage will need to be inspected, however, it is best to begin with the point of impact.
Answer C is wrong.
Answer D is wrong.

Question #153
Answer A is wrong. Always listen to customer's complaints and assume that the complaint is justified.
Answer B is wrong. Always listen to customer's complaints and assume that the complaint is justified.
Answer C is wrong.
Answer D is correct.

Question #154
Answer A is wrong. Not all parts are available as re-manufactured.
Answer B is correct. The supplier should be contacted to verify availability.
Answer C is wrong.
Answer D is wrong.

Question #155
Answer A is wrong. A is right.
Answer B is wrong. B is right.
Answer C is correct.
Answer D is wrong.

Question #156
Answer A is correct. Minor bends on a bumper shock mounting plate can be repaired.
Answer B is wrong. Minor bends can be repaired, the bumper shock is still strong.
Answer C is wrong.
Answer D is wrong.

Question #157
Answer A is wrong. A is right.
Answer B is wrong. B is right.
Answer C is correct.
Answer D is wrong.

Question #158
Answer A is wrong. A hood is a non-structural part.
Answer B is correct. A rocker panel is a structural part.
Answer C is wrong. A door is a non-structural part.
Answer D is wrong. A quarter panel is a non-structural part.

Question #159
Answer A is wrong. An evaporator is part of an A/C system.
Answer B is wrong. A condenser is part of an A/C system.
Answer C is wrong. An expansion valve is part of an A/C system.
Answer D is correct. A heater core is not part of an A/C system.

Question #160
Answer A is correct. The right rear is on the passenger's side.
Answer B is wrong. The right rear is on the passenger's side.
Answer C is wrong.
Answer D is wrong.

Question #161
Answer A is wrong. A is right.
Answer B is wrong. B is right.
Answer C is correct.
Answer D is wrong.

Question #162
Answer A is wrong. Vehicle weight is part of the amount of inertia damage.
Answer B is wrong. Vehicle height is not part of inertia damage.
Answer C is correct. Vehicle speed is part of the amount of inertia damage.
Answer D is wrong. Vehicle occupant weight is part of the amount of inertia damage.

Question #163
Answer A is wrong.
Answer B is wrong.
Answer C is correct. The letters SUV stand for sport utility vehicle.
Answer D is wrong.

Question #164
Answer A is wrong.
Answer B is wrong.
Answer C is correct. An addition is a necessary non-included operation.
Answer D is wrong.

Question #165
Answer A is wrong.
Answer B is correct. A two-channel ABS controls the rear wheels independently.
Answer C is wrong.
Answer D is wrong.

Question #166
Answer A is wrong. Ultra high strength steel is not the most common type of steel on a vehicle.
Answer B is correct. Mild steel is the most common type of steel on a vehicle.
Answer C is wrong. High strength steel is not most common type of steel on a vehicle.
Answer D is wrong. Martensetic steel is a type of ultra high strength steel.

Question #167
Answer A is wrong.
Answer B is wrong.
Answer C is wrong.
Answer D is correct. The drawing is a master cylinder and power booster.

Question #168
Answer A is wrong. Collision estimating guides do not include add-on accessories.
Answer B is correct. Collision estimating guides do not include add-on accessories.
Answer C is wrong.
Answer D is wrong.

Question #169
Answer A is correct. Safety components protect the passengers.
Answer B is wrong. Safety components protect the passengers.
Answer C is wrong. Safety components protect the passengers.
Answer D is wrong. Safety components protect the passengers.

Question #170
Answer A is correct. Bolts are fasteners and should be included in additional materials.
Answer B is wrong. Name plates should be listed as a separate item.
Answer C is wrong.
Answer D is wrong.

Question #171
Answer A is wrong. Imported vehicles are broken down into region, Asian or European.
Answer B is correct. The collision estimating guide should determine estimate sequence.
Answer C is wrong.
Answer D is wrong.

Question #172
Answer A is wrong. A is right.
Answer B is wrong. B is right.
Answer C is correct.
Answer D is wrong.

Question #173
Answer A is wrong. Reconditioned metal bumpers are called rechromed bumpers.
Answer B is wrong. Reconditioned metal bumpers are called rechromed bumpers.
Answer C is correct. Reconditioned metal bumpers are called rechromed bumpers.
Answer D is wrong. Reconditioned metal bumpers are called rechromed bumpers.

Question #174
Answer A is correct. Flexible plastics may be reshaped with heat.
Answer B is wrong. Not all types of plastic may be welded.
Answer C is wrong.
Answer D is wrong.

Question #175
Answer A is wrong. 66635 is the number of the right-hand part.
Answer B is correct. 66635 is the number of the right-hand part.
Answer C is wrong.
Answer D is wrong.

Question #176
Answer A is wrong. Seam sealer application is not part of new replacement part corrosion protection.
Answer B is wrong. A weld area is a corrosion or rust hot spot.
Answer C is wrong.
Answer D is correct.

Question #177
Answer A is wrong. A is right.
Answer B is wrong. B is right.
Answer C is correct.
Answer D is wrong.

Question #178
Answer A is wrong. A is right.
Answer B is wrong. B is right.
Answer C is correct.
Answer D is wrong.

Question #179
Answer A is wrong. A is right.
Answer B is wrong. B is right.
Answer C is correct.
Answer D is wrong.

Question #180
Answer A is wrong. Estimator A is correct.
Answer B is wrong. Estimator B is correct.
Answer C is correct.
Answer D is wrong.

Question #181
Answer A is wrong. The customer will not understand technical jargon.
Answer B is correct. the customer will understand layman's language.
Answer C is wrong.
Answer D is wrong.

Question #182
Answer A is wrong. Judgment time includes rough out.
Answer B is correct. Judgment time does not include refinish.
Answer C is wrong. Judgment time includes damage analysis.
Answer D is wrong. Judgment time includes plan repair.

Question #183
Answer A is correct. A rear clip top includes a slice weld of the rocker panels.
Answer B is wrong. The rear frame rails are not spliced in a rear clip top.
Answer C is wrong.
Answer D is wrong.

Question #184
Answer A is wrong. Many times damaged panels can be repaired, rather than replaced.
Answer B is correct. If the panel repair cost exceeds the replacement cost, the panel should be replaced.
Answer C is wrong.
Answer D is wrong.

Question #185
Answer A is wrong. A is right.
Answer B is wrong. B is right.
Answer C is correct.
Answer D is wrong.

Question #186
Answer A is wrong. A is right.
Answer B is wrong. B is right.
Answer C is correct.
Answer D is wrong.

Question #187
Answer A is wrong. A is right.
Answer B is wrong. B is right.
Answer C is correct.
Answer D is wrong.

Question #188
Answer A is correct. Fenders are not re-manufactured.
Answer B is wrong. Wheels are re-manufactured.
Answer C is wrong. Bumpers are re-manufactured.
Answer D is wrong. Alternators are re-manufactured.

Question #189
Answer A is wrong. A is right.
Answer B is wrong. B is right.
Answer C is correct.
Answer D is wrong.

Question #190
Answer A is wrong. The part placement on a vehicle has no affect on the repair/replace decisions.
Answer B is correct. The cost of the repair versus cost of replacement must be considered.
Answer C is wrong. The color of the part has no affect on the repair/replace decisions.
Answer D is wrong. The composition of the part has no affect on the repair/replace decisions.

Question #191
Answer A is wrong. Aftermarket part labor time is not listed in the collision estimating guide.
Answer B is wrong. Aftermarket part prices are not listed in the collision estimating guides.
Answer C is wrong.
Answer D is correct.

Question #192
Answer A is wrong. A is right.
Answer B is wrong. B is right.
Answer C is correct.
Answer D is wrong.

Question #193
Answer A is wrong. The customer will understand the length of repair time if the repair steps are explained.
Answer B is correct. The customer will understand the length of repair time if explained.
Answer C is wrong. The customer will understand the length of repair time if explained.
Answer D is wrong. The customer will understand the length of repair time if explained.

Question #194
Answer A is wrong. A is right.
Answer B is wrong. B is right.
Answer C is correct.
Answer D is wrong.

Question #195
Answer A is wrong. Estimator A is correct.
Answer B is wrong. Estimator B is correct.
Answer C is correct.
Answer D is wrong.

Question #196
Answer A is wrong. 2.5 x 0.4 = 1.0
Answer B is correct. 2.5 x 0.4 = 1.0
Answer C is wrong. 2.5 x 0.4 = 1.0
Answer D is wrong. 2.5 x 0.4 = 1.0

Question #197
Answer A is wrong. A is right.
Answer B is wrong. B is right.
Answer C is correct.
Answer D is wrong.

Question #198
Answer A is wrong. A is right.
Answer B is wrong. B is right.
Answer C is correct.
Answer D is wrong.

Question #199
Answer A is wrong. A is right.
Answer B is wrong. B is right.
Answer C is correct.
Answer D is wrong.

Question #200
Answer A is wrong. Severed fuel pump wiring could cause a no-start.
Answer B is wrong. An inertia switch could cause a no-start.
Answer C is correct. A severed knock sensor will not cause a no-start.
Answer D is wrong. A disconnected spark and fuel sensor could cause a no-start.

Question #201
Answer A is wrong. Alternators are available as aftermarket.
Answer B is correct. Air bag modules are not available as aftermarket.
Answer C is wrong. Idler arms are available as aftermarket.
Answer D is wrong. Fenders are available as aftermarket.

Question #202
Answer A is correct. Because aftermarket parts cost less, fewer repairs are made. The damaged parts are replaced with aftermarket parts.
Answer B is wrong. Aftermarket parts cost less than OEM.
Answer C is wrong.
Answer D is wrong.

Question #203
Answer A is wrong. A is right.
Answer B is wrong. B is right.
Answer C is correct.
Answer D is wrong. B

Question #204
Answer A is wrong. An alternator is pictured.
Answer B is wrong. An alternator is pictured.
Answer C is correct. An alternator is pictured.
Answer D is wrong. An alternator is pictured.

Question #205
Answer A is correct. Loss of strength is the most important aspect of structural damage
Answer B is wrong. Loss of structural part weight is not the most important aspect
Answer C is wrong. Loss of paint is not important.
Answer D is wrong. Loss of dimension is an important aspect, but not the most important aspect

Question #206
Answer A is wrong. Refinish time is included in head notes.
Answer B is correct. Judgment time is not included in head notes.
Answer C is wrong. R&I time is included in head notes.
Answer D is wrong. O/H time is included in head notes.

Question #207
Answer A is correct. Warranties are non-negotiable.
Answer B is wrong. Aftermarket part use is negotiable.
Answer C is wrong. Betterment is negotiable.
Answer D is wrong. Frame time is negotiable.

Question #208
Answer A is wrong. The seat belt should be replaced if it has cut webbing.
Answer B is wrong. The seat belt should be replaced if it has bowed webbing.
Answer C is correct. The seat belt need not be replaced, if it has a stain.
Answer D is wrong. The seat belt should be replaced if it has broken threads.

Question #209
Answer A is wrong. Deployed air bags can not be repacked.
Answer B is wrong. Undamaged crash sensors may not need to be replaced.
Answer C is wrong.
Answer D is correct.

Question #210
Answer A is correct. The full frame is stronger than a unibody or space frame.
Answer B is wrong. The unibody is not as strong as a full frame.
Answer C is wrong. The space frame, similar to a unibody, is not as strong as a full frame.
Answer D is wrong. Unit frame is not a frame type.

Glossary

A Abbreviation for ampere.

Abrasive A material such as sand, crushed steel grit aluminum oxide, silicon carbide, or crushed slag used for cleaning or surface roughening.

Abrasive coating (1) In closed coating paper, the complete surface of the paper is covered with abrasive; no adhesive is exposed. (2) In open coating, adhesive is exposed between the grains of abrasive.

Accent stripes Lines applied to a vehicle to add a decorative, customized look.

Access hole An opening that permits a technician to access fasteners and other components inside a door.

Accessible area An area that can be reached without parts being removed from the vehicle.

Accessories Items that are not essential to the operation of a vehicle, such as the cigarette lighter, radio, luggage rack, or heater.

Access time The time required to remove extensively damaged collision parts by cutting, pushing, or pulling.

Acetylene A gas used in flame welding and cutting.

Acid core A type of solder in tubular wire form having an acid flux paste core.

Acrylic A thermoplastic synthetic resin used in both emulsion and solvent-based paints, available as a lacquer or enamel.

Acrylic enamel A type of finish that contains polyurethane and acrylic additives.

Acrylic polyurethane enamel A material with great weatherability that generally provides higher gloss and greater durability than other polyurethane enamels.

Activator An additive used to cure a two- or multi-package enamel.

Active restraint A seat belt that the occupant of a vehicle must fasten.

Actual cash value (ACV) The current market value of a standard production vehicle and its accessory options as determined by used car guidebook listings or car dealer assessments.

Additive A chemical substance that is added to a finish, in small amounts, to impart or improve desirable properties.

Adhesion The ability of one substance to stick to another.

Adhesion promoter A water-white, ready-to-spray lacquer material that provides a chemical etch to original equipment manufacturer (OEM) finishes.

Adhesive backed molding A trim piece provided with an adhesive back coating to simplify installation.

Adhesive bonding A mechanical bonding between the adhesive and the surfaces that are being joined together.

Adhesive caulk A material used to seal or join seams, and used to install windshields and rear window glass.

Adhesive compound A nonhardening caulk-type material used to hold fixed glass in place.

Adhesive joining The assembly of components with a chemical bonding agent.

Adhesive primer/hardener A material brushed on a mirror support and on the glass before applying the mirror mounting adhesive.

Adjuster An insurance company representative, often called an appraiser, responsible for approving a collision repair bid to satisfy a vehicle damage claim.

Adjusting slot The elongated holes on mounting brackets and bumper shock absorbers that permit alignment during installation.

Adjustment The bringing of parts into alignment or proper dimensions with fasteners that permit some movement for position.

Aerodynamic A body shape having a low wind resistance.

Aftermarket Any parts or accessories, new or used, that are installed after original manufacture of the vehicle.

Aging The process of permitting a material to stand for a period of time.

Agitator A paint mixer or stirrer of any type.

Aiming screw The adjusting screws used to aim a headlight.

Air (1) A natural gas, usually used under pressure as a propellant. (2) The natural gas we breathe.

Air bag system A system that uses impact sensors, an on-board computer, an inflation module, and a nylon bag in the dash and/or steering column to protect the passenger and/or driver during a collision.

Air brush A small paint spray device used for fine detailing, fish scaling, and similar paint work.

Air cap The component located at the front of the gun that directs the compressed air into the material stream.

Air compressor Equipment that is used to supply pressurized air to operate shop tools and equipment.

Air conditioning system A system having a compressor, evaporator, condenser, and associated components that cool the air in the passenger compartment of a vehicle.

Air dam The structure mounted under the front portion of a vehicle designed to direct air through the radiator and across the engine.

Air drying The act of allowing a painted surface to dry at ambient temperature without the aid of an external source.

Air filter A device used to trap dirt particles or other debris in an air line.

Air gun A device that uses compressed air to clean and dry surfaces and areas.

Airless spraying A method of spray paint application in which atomization is effected by forcing paint, under high pressure, through a very small orifice in a spray gun cap.

Air line A pipe or flexible hose used to transport compressed air from one point to another.

Air make-up system A method used to replace the air that is exhausted from a paint spray booth.

Air-over-hydraulic system A system that utilizes a pneumatic motor to drive a hydraulic pump.

Air pressure (1) The force exerted on a container by compressed air. (2) The pressure of the surrounding ambient air.

Air-purifying respirator A filtered breathing aid used to clean or purify air.

Air spray A method of applying paint in the form of tiny droplets in air as paint is atomized by a spray gun as a result of being forced into a high velocity air stream.

Air supply system An air pumping system to supply fresh air for breathing to a painter in a paint spray booth.

Air suspension A vehicle suspension system that makes use of pneumatic cylinders to replace or supplement mechanical springs.

Air transformer A pneumatic control device that is used to filter and control the air delivered by a compressor.

Alcohol A colorless volatile liquid (1) used as diluents, solvents, or co-solvents in paints and (2) used as a fuel for racing engines.

Align To make an adjustment to a line or to a predetermined relative position.

Alignment The arrangement of a vehicle's basic structural components in relation to each other.

Alignment gap The space between two components, such as a door and pillar or fender and hood.

Alkyd A chemical combination of an alcohol, an acid, and an oil useful in water-based house paint and automotive primers.

Alkyd enamel The least expensive of the enamels.

Alligatoring A paint finish defect that resembles the pattern of an alligator's skin.

Alloy A mixture of two or more metals.

Alternating current Electrical current that changes direction.

Alternator An electrical device that is used to generate alternating current, which is then rectified into direct current for use in the vehicle's electrical system.

Aluminum (Al) (1) A lightweight metal. (2) A material that is useful as a substrate or pigment.

Aluminum oxide An extremely tough abrasive that is highly resistant to fracturing and capable of penetrating hard surfaces without dulling.

Ambient temperature The temperature of the surrounding air.

Ammeter An electrical instrument that is used to measure electrical current.

Amp An abbreviation for ampere.

Ampere (A) An electrical unit for current.

Anchor To hold in place.

Anodizing An electrolytic surface treatment for aluminum that builds up an aluminum oxide coating.

Antenna wire circuit A circuit placed between layers of glass or printed on the interior surface of the glass.

Antifouling A paint containing toxic substances that inhibit the growth of certain organisms on ship bottoms.

Anti-lacerative glass A form of laminated glass having an additional layer of plastic on the occupant side to contain fragments of glass during breakage.

A-pillar The windshield post.

Appraiser An insurance company representative, often referred to as an adjuster, responsible for approving a collision repair bid.

Apprentice A person who learns his or her trade while working under the supervision of a skilled technician concurrent with classroom training.

Arc spot welding The process of making spot welds with the heat of an arc, primarily in areas that are not accessible with resistance spot welders.

Arc welding A joining technique that uses an electrical arc as the heat source.

Asbestos A cancer-causing material that was once used in the manufacture of brake and clutch linings.

Asbestos dust A cancer-causing by-product agent of a material that was once used in the manufacture of brake and clutch lining assemblies.

ASE certification A voluntary testing program to help give recognition that you are a knowledgeable collision repair technician, estimator, or painter.

Asphyxiation The inability to breathe due to anything that prevents normal breathing, such as mists, gases, and fumes.

Assembly Two or more parts that are bolted or welded together to form a single unit.

Assembly drawing Drawings used by the manufacturer when assembling a component or a vehicle.

Atmosphere-supplied respirator A system that has a blower or a special compressor to supply clean fresh outside air to a face mask or hood.

Atomize To break a liquid into a fine mist.

Auto-ignition temperature The approximate lowest temperature at which a flammable gas or vapor-air mixture will ignite without spark or flame.

Automatic cut-off A safety device used to shut off the air compressor at a preset pressure.

Automatic welding Welding equipment that performs the welding operation without constant observation and adjustment of the controls.

Average retail price The local value of a like vehicle, based on actual sales reports from new and used car dealers.

Backfire (1) A small explosion that can cause a sharp popping sound in oxygen-acetylene welding equipment. (2) An explosion in the exhaust system of a motor vehicle. (3) An explosion in the intake manifold of a vehicle.

Backlight A vehicle window located behind the occupants.

Baffle A panel used to direct air to the radiator.

Baked enamel A finish achieved when heat is used to promote rapid drying.

Baked-on finish A painted surface that has been cured by heating after application.

Banding A single coat applied in a small spray pattern to frame in an area to be painted.

Bar gauge A gauge used to accurately measure and diagnose body and frame collision damages for all conventional and unitized vehicles.

Barrier cream A hand cream that provides protection and soothes the skin when working with irritating materials.

Base The resin component of paint to which color pigments and other components are added.

Basecoat The coat of paint on which final coats are applied.

Basecoat/clearcoat A paint method in which the color effect is given by a highly pigmented basecoat followed by a clearcoat to provide gloss and durability.

Base metal Any metal to be welded or cut.

Bathtub clip Small plastic attachment pieces that are pressed into holes in the vehicle body to which moldings are attached.

Battery charger An electrical device used to recharge a vehicle battery.

Bead The amount of filler metal or plastic material deposited in a joint when welding two pieces together.

Bearing plate A plastic spacer used to reduce the friction between the door handle and the trim panel.

Belt line A horizontal molding or crown along the side of the vehicle at the bottom of the glass.

Belt trim support retainers The parts that hold the window glass in position and prevent it from wobbling.

Bench A vehicle underbody anchoring device for checking frame and suspension dimensions for damage and to allow straightening procedures.

Bench grinder A tool used for sharpening or metal removal that is bolted to a workbench and driven by an electric motor.

Bench system An alignment method using equipment that allows the vehicle to be set on preadjusted pins to check for damage.

Bench vise A heavy adjustable bench mounted holding tool.

Bend A change in a body or trim part from its original shape.

Bezel A trim ring that surrounds a headlight or gauge.

Binder An ingredient in paint that holds pigment particles together.

Bleeder gun A spray gun design that allows air to pass through the gun at all times, preventing pressure buildup in the supply lines.

Bleeding The original color showing through after a new coat has been applied.

Blending (1) The mixing of two or more paint colors to achieve the desired color. (2) The technique used in repairing acrylic lacquer finishes, extending each color coat a little beyond the previous coat to blend into surrounding finish.

Blistering The formation of hollow bubbles or water droplets in a paint film, usually caused by expansion of air or moisture trapped beneath the paint film.

Block sanding The use of a flat object, such as a block of wood, and sandpaper to obtain a flat surface.

Bloom A clouded appearance on a finish paint coat.

Blowgun A device that uses blasts of compressed air to help clean and dry work surfaces.

Blushing The hazing or whitening of a film caused by absorption and retention of moisture in the drying paint film.

Bodied cement A syrupy solvent cement that is composed of a solvent and a small quantity of compatible plastic.

Body (1) The consistency of a liquid; the apparent viscosity of a paint as assessed when stirring it. (2) An assembly of sheet metal parts that comprise the enclosure of a vehicle.

Body code plate A manufacturer-mounted plate on a vehicle providing body type and other information.

Body file A flat, half-round, or round file designed to be pushed across the work surface in the direction of the cutting teeth.

Body filler A heavy-bodied plastic-like material that cures very hard; used to fill small dents in metal.

Body hammer A specialized hammer used to reshape damaged sheet metal.

Body hardware Appearance and functional parts, such as handles, on the interior and exterior of a vehicle.

Body mounting The method and means by which a vehicle body is placed onto a chassis.

Body-over-frame A vehicle having separate body and chassis parts bolted to the frame.

Body-over-frame construction A method in which the automobile body is bolted to a separate frame.

Body panel A sheet of shaped steel, aluminum, or plastic that forms part of the car body.

Body saw A saw equipped with an abrasive blade used to cut floor sections or panels.

Body solder An alloy of tin and lead, used to fill dents and other body defects.

Body technician A person who performs such basic repair tasks as removing dents, replacing damaged parts, welding metal, filling, sectioning, and sanding.

Body trim (1) The material used to finish the interior of the passenger and trunk compartments. (2) The rubber and metal moldings on the exterior of a vehicle.

Bolt-through glass Glass that is held to the regulator mechanism with mechanical fasteners, such as bolts or rivets.

Bonding strips Strips of aluminum, fiberglass, or aluminum and stainless steel tape, used to patch holes in vehicle bodies.

Bounce-back The atomized particles of paint that rebound from the surface being sprayed and contribute to overspray.

B-pillar The pillar between the belt line and roof between the front and rear doors on four-door vehicles and station wagons.

Brackets A part that is used to attach components to each other or to the body and frame.

Braze welding A method of welding using a filler metal.

Bridging The characteristic of an undercoat that occurs when a scratch or surface imperfection is not completely filled.

Brittle A lack of toughness and flexibility.

Bronzing The formation of a metallic-appearing haze on a paint film.

Brush (1) A method of applying paint. (2) An applicator for applying paint.

Brushing The act of applying paint using a brush.

Buckles The distortion, ridges, or high places on a metal body part as a result of collision damage.

Buffer A tool that resembles a disk sander, but runs at a higher speed and uses a polishing bonnet, to buff the final coat of paint.

Buffing compound An abrasive paste or cake used with a cloth or sheepskin pad to remove fine scratches and to polish lacquer finishes.

Build The amount of paint film deposited, measured in mils.

Bulge A high crown or area of stretched metal.

Bumping The process of smoothing the damaged area following roughing.

Bumping file A tool with a spoon-like shape and serrated surface, used to slap down and shrink high spots.

Bumping hammer A hammer used to roughly pound out a dent.

Bumping spoon A spoon that is often used as a pry bar.

Burnishing The polishing or buffing of a finish by hand or machine using a compound or liquid.

Butt weld A weld made along a line where two pieces of metal are placed edge to edge.

Butyl acetate A solvent for paint, often used in lacquers.

Butyl adhesive A rubber-like compound used to bond fixed glass in place.

Butyl tape Tape that is used with a separate adhesive to bond fixed glass in place.

Caged plate assembly A tapped plate inside a sheet metal box spot welded to the inside of a door or pillar to accept the hinge and striker bolts; is moved in the cage for adjustment.

Calcium A metal component of dryers and pigments.

Calibrate The process of checking equipment to see that it meets test specifications, and that the settings of the equipment are correct.

Camber The inward or outward tilt of a wheel at top from true vertical.

Canister (1) A container of chemicals designed to remove specific vapors and gases from breathing air. (2) A container filled with charcoal used to filter and trap fuel vapors.

Carbonizing flame A welding flame with excess acetylene, which will introduce carbon into the molten metal.

Carbon monoxide (CO) A deadly gas created by the incomplete burning of fuel.

Carnauba wax A hard wax obtained from a species of palm tree and used in some body polishing materials.

Carpet The floor covering used in vehicles.

Caster The backward or forward tilt of a kingpin or spindle support arm at top from true vertical.

Casting (1) The process of molding materials using only atmospheric pressure. (2) A part produced by that process.

Catalyst A substance that causes or speeds up a chemical reaction when mixed with another substance but does not change by itself.

Catalytic converter An exhaust component used to reduce harmful exhaust gas emission through a chemical reaction.

Caulking compound A semiflexible sealer used to fill cracks, seams, joints, and to eliminate water or air leaks and rattles.

C-clamp A C-shaped device with threaded posts used to hold parts together during assembly or welding procedures.

Centering gauge A frame gauge, used in sets of four to locate the horizontal datum planes and centerline on a vehicle.

Centerline strut gauge A device used to detect misalignment of strut towers in relation to the center plane and the datum plane.

Center pillar A box-like column that separates the front and rear doors on a four-door vehicle. Often called a hinge pillar, it holds the front door striker/lock and the rear door hinges.

Center plane An imaginary centerline running lengthwise along the datum plane.

Center punch A tool having a pointed, tapered shaft, used to locate and make starting dents for drilling.

Centimeter A unit of linear measure in the SI metric system.

Ceramic mask An opaque black mask baked directly on the perimeter of the glass to help hide the black polyurethane adhesive and the butyl tape.

Cerium oxide powder A very fine abrasive used to polish out scratches on glass.

Chalking The result of weathering of a paint film characterized by loose pigment particles on the surface of the paint.

Chassis An assembly of mechanisms that make up a major operating system of a vehicle, including everything under the body—suspension system, brake system, wheels, and steering system.

Check A small crack.

Checking A type of failure in which cracks in the paint film begin at the surface and progress downward, usually resulting in a straight V-shaped crack, narrower at the bottom than at the top.

Check valve A device that only allows the passage of air or fluid in one direction.

Chemical burn An injury that results when a corrosive chemical strikes the skin or eye.

Chemical shining A spotty discoloration of the topcoat caused by atmospheric conditions, often occurring near an industrial area.

Chipping The breaking away of a small portion of paint film because of its inability to flex under impact or thermal expansion and contraction.

Chlorofluorocarbon A class of refrigerant chemicals that have a detrimental effect on the earth's ozone layer.

Chopping A term that describes the lowering of a vehicle profile.

Chroma The intensity of a color; the degree it differs from the white, gray, or black of the neutral axis of the color tree.

Chronic effect The adverse effect on a human or animal having symptoms that develop slowly over a long period of time or that may reoccur frequently.

Circuit A path for electricity. A circuit must be complete before current will flow.

Circuit breaker A device having a bimetallic strip and a set of contacts that open if excessive current heats the strip to make it bend.

Claimant A person who files a claim.

Clamp saw A saw that is used to cut through spot welds by only cutting through the top layer of material.

Clean To be free of dirt or other material, such as after washing. A bright clear color.

Cleaner A material that is used to clean a substrate.

Clear Transparent. A paint that does not contain pigment or only contains transparent pigment.

Clearance The amount of space between adjacent parts or panels.

Clearcoat A transparent top coating on a painted surface so the base color coat is visible.

Clip A section of a salvaged vehicle. A mechanical fastener used to hold moldings to panel.

Clip removal tool A hand tool used for removing the mechanical fasteners used to attach door handles and trim.

Closed bid One in which the final cost has been determined.

Closed coat abrasive A material in which the abrasive grains are as close together as possible.

Closed structural member A boxed-in section, generally only accessible from the outside, such as a rail or a pillar.

Clouding The formation or the presence of a haze in a liquid or on a film.

Coat, double Two single coats of material, one followed by the other with little or no flash time.

Coated abrasive A combination of abrasive grains, backing materials, and bonding agents (sandpaper, for example).

Coating The act of applying paint; the actual film left on a substrate by the paint.

Coatings Covering material used to protect an area.

Coat, single A coat of material produced by two passes of a spray gun, one pass overlapping the other in half steps.

Cobwebbing The tendency of sprayed paint to form strings or strands rather than droplets as it leaves the spray gun.

Cohesive bonding A process of joining plastics that involves using solvent cements to melt plastic materials together.

Cold knife A device used to cut through adhesive that secures fixed glass or other components.

Cold shrinking The reducing of a surface area of metal by using a shrinking hammer and shrinking dolly.

Cold working The working of metal without the application of heat.

Collision A "wreck" or "crash." Damage caused by an impact on a vehicle body and chassis.

Color The visual appearance of a material, such as red, yellow, blue, or green.

Color holdout The ability of a primer-sealer to allow the finish coat to maintain its high gloss.

Color retention The permanence of a color under a specific set of conditions.

Color sanding The process of wet sanding to remove surface imperfections from the topcoat.

Commission The payment to a technician on the basis of a percentage, usually 40 to 60 percent, of the total labor for a repair job.

Compatibility The ability of two or more materials to be used together.

Composite A plastic formed by combining two or more materials, such as a polymer resin (matrix) and a reinforcement material.

Composite headlight A headlight with a separate bulb, lens, and connector.

Compounding The action of using an abrasive material either by hand or by machine to smooth and bring out the applied topcoat.

Compression resistance spot welding A method using clamping force and resistance heating to form two-sided spot welds.

Compressor A device used to deliver compressed air for the operation of shop equipment. A device used to circulate refrigerant in an air conditioning system.

Computer An electronic device for storing, manipulating, and disseminating information.

Concave An inward curve, like a dent.

Concentration The amount or ratio of any substance in a solution.

Condensation A change in state of a vapor to a liquid on a cold surface, usually moisture. A type of polymerization characterized by the reaction of two or more monomers to form a polymer plus some other product, usually water.

Conductor A material that will allow the flow of electrons.

Connector A device having male and female halves which are fastened together to join a wire or wires.

Consistency The fluidity of a liquid or of a system.

Contaminants The foreign substances on a prepared surface to be painted, or in the paint, that will adversely affect the finish. Any impurities in a material.

Continuity tester A device that is used to test bulbs, fuses, switches, circuits, and other electrical devices.

Continuous sander A sander that uses an abrasive belt to remove paint and to sand body filler.

Contract An agreement in the form of a written legal document between two or more people.

Control point The points on a vehicle, including holes, flats, or other identifying areas, used to position panels and rails during the manufacture of the vehicle.

Conventional frame A vehicle construction type in which the engine and body are bolted to a separate frame.

Conventional point The points on a unibody used as a reference to make a repair.

Conversion coating A chemical treatment used on galvanized steel, uncoated steel, and aluminum to prevent rust.

Convertible top fabric Canvas, synthetic fabric, and vinyl cloth material.

Convex An outward curve, like a bump.

Copper (Cu) A metal. A difficult metal substrate to paint. A metal used in the manufacture of pigments and dryers.

Core (1) The tubes that form the coolant passages in a radiator or heat exchanger. (2) A rebuildable unit used for exchange when purchasing a new or rebuilt unit.

Corporate Average Fuel Economy (CAFE) The goal of 40 miles per gallon of fuel is to be accomplished by the year 2000 by every automobile manufacturer.

Corporation A business that generally has two or more owners.

Corrosion A chemical reaction caused by air, moisture, or corrosive materials on a metal surface, referred to as rusting or oxidation.

Corrosion protection The use of any of a variety of methods to protect steel body parts from corrosion and rusting.

Courtesy estimate A written estimate for record-keeping purposes.

Coverage The area that a given amount of paint will cover when applied according to the manufacturer's directions.

Cowl cut A nose or front end that is cut behind the cowl or firewall.

Cowl panel A panel forward of the passenger compartment to which the fenders, hood, and dashboard are bolted.

C-pillar The pillar connecting the roof to the rear quarter panel.

Cracking The splitting of a paint film, usually as straight lines, which penetrate the film thickness, often the result of overbaking.

Crane A portable device that is used to lift or move heavy objects, such as a vehicle engine.

Cratering The formation of holes in the paint film where paint fails to cover due to surface contamination.

Crawling A wet paint film defect that results in the paint film pulling away from, or not wetting, certain areas.

Crazing A paint film failure that results in surface distortion or fine cracking.

Creeper A low platform having four caster rollers used by a technician for mobility when working under a vehicle.

Creeping A condition in which paint seeps under the masking tape.

Cross member A reinforcing piece that connects the side rails of a vehicle frame.

Crown A convex curve or line in a body panel.

Crush zone A section built into the frame or body designed to collapse and absorb some of the energy of a collision.

Cubic centimeter A unit measure of volume in the SI metric system equal to one milliliter.

Cure The process of drying or hardening of a paint film.

Curing The chemical reaction in drying of paints that dry by chemical change.

Current The flow of electrons in an electrical circuit.

Customer requested Repairs that the customer wishes to have made, outside those covered by insurance.

Customize The altering of a vehicle to meet individual tastes or specifications.

Custom-mixed topcoat A special blend of paint that is often used when attempting to match a badly oxidized or faded finish.

Custom paint The refinishing or decorating of a vehicle in a personalized manner.

Cyanoacrylate adhesive system A two-part adhesive that forms an extremely strong bond on hard plastics and similar materials.

Datum line An imaginary line that appears on frame blueprints or charts to help determine the correct height.

Datum plane The horizontal plane or line used to determine correct frame measurements.

Decal Paint films in the form of pictures or letters that can be transferred from a temporary paper backing to another surface.

Dedicated fixture measuring system A bench with fixtures that may be placed in specific points for body measurement.

Deductible The amount of the claim that the vehicle owner must pay, with the remainder paid by the insurance company.

Defroster circuit Narrow bands of a conductive coating that are printed on the interior surface of the rear window glass.

Degradation The gradual or rapid disintegration of a paint film.

Degreasing The cleaning of a substrate, usually metal, by removing greases, oils, and other surface contaminants.

Dehydration The removal of water.

Density The weight of a material per unit of volume, usually grams per cubic centimeter.

Depreciation The loss of value of a vehicle or other property due to age, wear, or damage.

Depression A concave dent.

Detailing The final cleanup and touch-up on a vehicle.

Dew point The temperature at which water vapor condenses from the air.

Diamond A damage condition where one side of the vehicle has been moved to the rear or the front, causing the frame/body to be out of square, or diamond-shaped.

Die A cutting tool used to make external threads or to restore damaged threads.

Diluent A liquid, not a true solvent, used to lower the cost of a paint thinner system.

Dilution ratio The amount of a diluent that can be added to any true solvent when the mixture is used to dissolve a certain weight of polymer.

Dinging The process of using a body hammer and a dolly to remove minor dents.

Dinging hammer A hammer used for removing dents.

Dipping Applying paint, primer, or sealer by immersing the part in a container of paint, withdrawing it, and then allowing the excess to drain.

Direct current (DC) Electrical current that travels in only one direction.

Direct damage Damage that occurs to an area that is in direct contact with the damaging force or impact.

Dirty (1) A color that is not bright and clean, that appears grayish. (2) A condition that requires cleaning.

Disassemble To take apart.

Disc grinder An electric or pneumatic tool used with abrasive wheels or discs for heavy-duty grinding, deburring, and for smoothing welds.

Disc sander A rotary power tool used to remove paint and locate low spots in a panel.

Dispersion The act of distributing solid particles uniformly throughout a liquid; commonly, dispersion of pigment in a vehicle.

Dispersion coatings Types of paint in which the binder molecules are present as colloidal particles instead of solutions.

Distillation range The boiling temperature range of a liquid.

Distortion A condition in which a component is bent, twisted, or stretched out of its original shape.

Doghouse The front clip or front body section, also called the nose section, including everything between the front bumper and the firewall.

Dog tracking The off-center tracking of the rear wheels as related to the front wheels.

DOL An abbreviation for the U.S. Department of Labor.

Dolly block An anvil-like metal hand tool held on one side of a dented panel while the other side is struck with a dinging hammer.

Domestic The classification for any vehicle having 75 percent or more of its parts manufactured in the United States.

Door hinge wrench A specially shaped wrench used to remove or install the retaining bolts on door hinges.

Door lock assembly The assembly that makes up a door lock.

Door lock striker That part of a door lock assembly that is engaged by the latching mechanism when the door is closed.

Door skin The outer door panel.

Double coat The technique of spraying the first pass left to right and spraying the second pass right to left directly over the first pattern.

Drain hole A hole in the bottom of a door, rocker panel, or other component that allows water to drain.

Drier (1) A chemical added to paint to reduce drying or cure time. (2) A system of removing moisture.

Drifting (1) The mixing of two or more colors to achieve the desired final color. (2) A term used for blending.

Drift punch A tool that has a long tapered shaft with a flat tip, used to align holes.

Drill A power tool with interchangeable cutting bits that are used to bore holes.

Drill bit A cutting tool used with a drill.

Drill press A floor- or bench-mounted electric drill used to bore holes.

Drip cap A term used for drip molding.

Drip molding The metal molding that serves as a rain gutter over doors; sometimes called a drip cap.

Drivetrain The engine, transmission, and drive shaft/axle assembly.

Drop light A portable light attached to an electrical cord used to illuminate a work area.

Dry (1) To change from a liquid to a solid which takes place after a paint is deposited on a surface. (2) To be free of moisture or other liquid.

Dryer (1) A catalyst added to a paint to speed up curing or drying time. (2) A device used for drying material.

Drying The process of changing a coat of paint from a liquid to a solid state due to evaporation of the solvent, a chemical reaction of the binding medium, or a combination of these causes.

Drying time The amount of time required to cure paint or allow solvents to evaporate.

Dry sanding A technique used to sand finishes without use of a liquid.

Dry spray An imperfect coat of paint, usually caused by spraying too far from the surface being painted or on too hot of a surface.

Dual-action sander A sander that combines circular and orbital motion in one device.

Ductility The property that permits metal deformation under tension.

Durability The length of service life; usually applies to a paint used for exterior purposes.

Early-model A term used to describe automobiles that are over fifteen years old.

Eccentric An offset cam-like section on a shaft used to convert rotary to reciprocating motion.

E-coat A cathodic electro-deposition coating process that produces a tough epoxy primer.

Edge joint A joint between the edges of two or more parallel or nearly parallel members.

Elastic deformation A condition that occurs when a material is not stretched beyond its elastic limit.

Elasticity A material's ability to be stretched and then return to its original shape.

Elastic limit The point at which a material will not return to its original shape after being stretched.

Elastomer A man-made compound with flexible and elastic properties.

Electrical system The starter, alternator, computers, wires, switches, sensors, fuses, circuit breakers, and lamps used in a vehicle.

Electric convertible top A convertible top that uses an electric motor and actuator assembly for raising and lowering.

Electric-over-hydraulic system A system having an electric motor to drive a hydraulic pump.

Electric tool A tool that operates on electrical power.

Electrocution A condition whereby electricity passes through the human body causing severe injury or death.

Electrode A metal rod used in arc welding that melts to help join the pieces to be welded.

Electronic display A light-emitting diode (LED), digital readout, or other device used to provide vehicle information.

Electronic system A computerized vehicle control system such as the engine control systems or antilock brake systems.

Electrostatic spraying The application of paint by high-voltage atomization.

Emulsion A suspension of fine polymer particles in a liquid, usually water.

Enamel A type of paint that dries in two stages: first by evaporation of the solvent and then by oxidation of the binder.

Endless line A custom painting technique in which narrow tape is applied in the desired design.

Energy reserve module An alternate source of power for an air bag system if the battery voltage is lost during a collision.

Entrepreneur One who has his or her own business.

Epoxy A class of resins that may be characterized by their good chemical resistance.

Epoxy fiberglass filler A waterproof fiberglass reinforcement material used for minor rust repair.

Estimating The analyzing of damage and calculating the cost of repairs.

Etching The chemical removal of a layer of base metal to prepare a surface for painting.

Ethylene dichloride A chemical used as a solvent to cement plastic joints.

Ethylene glycol A chemical base that is used for permanent antifreeze.

Evaporation A change in state from a liquid to a gas.

Evaporation rate The speed at which a liquid evaporates.

Expansion tank A small tank used to hold excess fuel or coolant as it expands when heated.

Exposure limit The limit set to minimize occupational exposure to a hazardous substance.

Extended method The replacement of an adhesive material when installing new fixed glass.

Extender pigment An inert, colorless, semitransparent pigment used in paints to fortify and lower the price of pigment systems.

Extension rod An arm used to connect a lock cylinder to a locking mechanism on a door latch assembly.

Exterior The outside area.

Exterior door handle A device that permits the opening of a door from the outside.

Exterior lock A lock used on front doors and trunk lids or hatches.

Exterior trim piece The moldings and other components applied to the outside of a vehicle.

External-mix gun A paint spray gun that mixes and atomizes the air and paint outside the air cap.

Eye bath A device that is used to flood the eyes with water in case of accidental contact with a chemical or other hazardous material.

Face bar A bare bumper with no hardware attachments.

Face shield A device worn to protect the face and eyes from airborne hazards and chemical splashes.

Factory-mixed topcoats Paint that is carefully proportioned at the factory to achieve the desired color and match the original finish.

Factory specifications Specific measurements and other information used during original manufacture of a vehicle.

Fading The loss of color.

Fan The spray pattern of a paint spray gun.

Fanning The use of pressurized air through a paint spray gun to speed up the drying time of a finish.

Fan shroud The plastic or metal enclosure placed around an engine-driven fan to direct air and improve fan action.

Fatigue failure A metal failure resulting from repeated stress that alters the character of the metal so that it cracks.

Feather The tapered edge between a bare metal panel and the painted surface.

Featheredge The tapering of the edge of the damaged area with sandpaper or special solvent.

Featheredge splitting Stretch marks or cracks along the featheredge that occur during drying or shortly after the topcoat has been applied over a primed surface.

Feathering The act of using sandpaper to taper the paint surface around a damaged area, from the base metal to topcoat.

Fiberglass A material composed of fine-spun filaments of glass used as insulation, and for reinforcement of a resin binder when repairing vehicle bodies.

Fiberglass cloth A heavy woven reinforcement material that provides the greatest physical strength of all the fibrous mats.

Fibrous composites A material that is composed of fiber reinforcements in a resin base.

Fibrous mat A reinforcing material consisting of nondirectional strands of chopped glass held together by a resinous binder.

Fibrous pad A pad of resin fiber or fiberglass that is used on the inside of the hood and other areas to deaden sound and provide thermal insulation.

File A tool with hardened ridges or teeth cut across its surface used for removing and smoothing metal.

Filler A material that is used to fill a damaged area.

Filler metal The metal added when making a weld.

Filler panel A panel that is found between the bumper and the body.

Filler strip A strip included in windshield installation kits to be used in the antenna lead area.

Film A very thin continuous sheet of material, such as paint that forms a film on the surface to which it is applied.

Finish (1) A protective coat of paint. (2) To apply a paint or paint system.

Finish coat The final coat of finish material applied to a vehicle.

Finishing hammer A hand tool used to hammer metal.

Finish sanding (1) The last stage in hand sanding the old finish. (2) Sanding the primer-surfacer using 400 grit or finer sandpaper.

Fish-eye A paint surface depression in wet paint film usually caused by silicone contamination of the paint.

Fish-eye eliminator An additive that makes paint less likely to show fish-eye.

Fixed glass Glass, such as a rear window, that is not designed to move.

Fixed pricing Standard pricing for performing a service or repair that does not change.

Fixture An accessory for a dedicated measuring system designed to attach to a bench to fit reference points.

Flaking A paint failure noted by large pieces of paint separating from the substrate.

Flames A flame-like design produced by the use of stencils and paint.

Flash The first stage of drying where some of the solvents evaporate, which dulls the surface from an exceedingly high gloss to a normal gloss.

Flashback A condition in which the oxygen-acetylene mixture burns back into the body of the welding torch.

Flasher unit An electrical device used to flash the turn signal and hazard lights.

Flash point The temperature at which the vapor of a liquid will ignite when a spark is struck.

Flash time The time between coats or paint application and/or baking.

Flat The lack of a gloss or shine.

Flat boy A term often used for a speed file.

Flat chisel A tool designed for shearing steel, including removing bolts or rivets.

Flat rate A predetermined time allowed for a particular repair and the money charged to perform that repair based on a standard per hour shop fee.

Flattener An additive that reduces the gloss of a finishing material.

Flex-additive A material added to a topcoat to make it flexible.

Floor jack Portable equipment used to lift a vehicle.

Floor pan The main underbody assembly of a vehicle that forms the floor of the passenger compartment.

Flow (1) The leveling characteristics of a wet paint film. (2) The ability of a liquid to run evenly from a surface and to leave a smooth film behind.

Fluid adjustment screw A spray gun control that is used to regulate the amount of material passing through the fluid tip as the trigger is depressed.

Fluid control valve A manual spray gun adjustment used to determine the amount of paint coming from the gun.

Fluid needle A spray gun valve component that shuts off the flow of material.

Fluid tip A spray gun nozzle that meters the paint and directs it into the air stream.

Fog coat A paint coat following a wet coat in which mottling or streaking occurs.

Foreign A general classification for a vehicle that has less than 75 percent of its parts manufactured in the United States.

Frame The heavy metal structure that supports the auto body and other external components.

Frame alignment The act of straightening a frame to the original specifications.

Frame-and-panel straightener A portable or stationary hydraulically powered device used to repair damaged sheet metal and correct frame damage.

Framed door A design in which the door frame surrounds and supports the glass.

Frame gauge A gauge that may be hung from the car frame to check alignment.

Frame straightener A pneumatic- or hydraulic-powered device used to align and straighten a distorted frame or body.

Frame system A heavy frame assembly upon which the various attachments are mounted.

Free air capacity The actual amount of free air that is available at the compressor's working pressure.

Frisket paper An adhesive-backed paper used as stencil material.

Front body hinge pillar The pillar to which the front door hinges are attached.

Front-wheel drive A vehicle that has its drive wheels located on the front axle.

Frosting (1) The formation of a surface haze. (2) The defects in a drying paint film. (3) The freezing of surface moisture on a line or component.

Full body section A section repair to both rocker panels, windshield pillars, and floor pan that is required to join the undamaged front half of one vehicle to an undamaged rear half of another vehicle.

Full cut-out method The replacement of an adhesive material when installing new fixed glass.

Full frame The strong, thick steel structure that extends from the front to the rear of a vehicle.

Full wet coat A heavy application of finish used to thoroughly cover the substrate.

Fuse An electrical protective device with a soft metal element that will melt and open an electrical circuit if more than the rated amount of current flows through it.

Fuse block A panel-like holder for fuses and circuit breakers to the vehicle's electrical circuits.

Fusibility A measure of a material's ability to join another while in a liquid state.

Fusible link A specially designed wire joint that melts and opens the circuit if excessive current flows.

Fusion weld A joining operation that involves the melting of two pieces of metal together.

Galvanized metal Metal that is coated with zinc.

Gap The distance between two points.

Garnish molding A decorative or finish molding around the inside of glass.

Gasket (1) A rubber strip used to secure fixed glass on early-model vehicles. (2) A cork, rubber, or combination of the two used as a seal between two mating surfaces.

Gas metal arc cutting (GMAC) An arc cutting process used to sever metals by melting them with the heat of an arc between a continuous metal electrode and the work.

Gas metal arc welding (GMAW) Also called MIG welding, an arc welding process that produces coalescence of metals by heating them with an arc between a continuous filler metal electrode and the work.

Gauge (1) A measure of thickness of sheet metal. (2) A device used to indicate a system condition, such as pressure or temperature.

General-purpose file A flat, round, or half-round shape tool used to remove burrs and sharp edges from metal parts.

General purpose hammer A hammer used for striking tools or tasks other than shaping sheet metal.

General purpose tool A tool that is common to any shop that performs automotive service or repair.

Glass installer The technician responsible for windshield and door glass replacement.

Glass run channel A term used for window channel.

Glass spacer The rubber pieces used to position or align glass.

Glazing putty A paste-like material used to fill small surface pits or flaws.

Gloss The ability of a surface to reflect light as measured by determining the percentage of light reflected from a surface at certain angles.

Goggle Glasses having colored lenses or clear safety glass that protects the eyes from harmful radiation during welding and cutting operations.

Grain pattern The surface appearance and color variation of vinyl fabric.

Grater file A file used to shape body filler before it has completely hardened.

Gravity feed gun A spray gun into which paint is fed by gravity.

Greenhouse The passenger portion of a vehicle body.

Grille The decorative panel in front of the radiator.

Grit A measure of the size of the particles on sandpaper or disc.

Grommet A donut-shaped rubber device used to surround wires or hoses for protection where they pass through holes in sheet metal.

Ground-return system In metal-framed vehicles, the frame is part of the electrical circuit, so only one wire is needed to complete the circuit. Composite or plastic bodies require two wires.

Guide coat A reference coat of a different color often applied to a primer-surfacer to be sanded off to visually determine if the panel is straight.

Hacksaw A hand saw used to cut metal.

Hand rubbing compound A rubbing compound designed for manual use only.

Hardener A curing agent used in certain plastics and epoxies.

Hard-faced hammer A hammer used to strike tools or to bend or straighten metal parts.

Hard hat A metal or plastic headgear worn to help protect the head from abrasions, hot sparks, and chemical sprays.

Hardness The quality of a dry paint film that gives film resistance to surface damage or deformation.

Hardtop An automobile body style that does not have a roof-supporting center pillar.

Hardtop door A door design without a frame around the glass that rests against the top and sides of the door opening and the weather stripping.

Hardware (1) Computer, printers, hard drives, CD-ROM drives, and other computer equipment. (2) Hinges, hangers, and fasteners used in vehicle construction.

Harness The electrical wires and cables that are tied together as a unit.

Hazardous material Any material that can cause serious physical harm or pose a risk to the environment.

Hazardous substance Any hazardous material that poses a threat to waterways and the environment.

Hazardous waste Any material that can endanger human health if handled or disposed of improperly.

Haze The development of a cloud in a film or in a clear liquid.

Header bar The framework or inner construction that joins the upper sections of the windshield or the back glass and pillars to form the upper portion of the windshield or the back glass opening, and that reinforces the top panel.

Head liner A cloth or plastic material covering the roof area inside the passenger compartment.

Heat gun An electric hand-held tool that blows heated air to soften plastic parts or for speed drying.

Heat shrink The heating with a torch, then using a hammer and dolly to flatten a panel, and then quenching it with water to shrink the metal.

Hem flange A flange at the bottom of a door panel.

Hemming tool A pneumatic tool used to create a hem seam.

Hem seam A door bottom seam formed by bending the outer panel hem flange over the inner panel flange with a hammer and a dolly.

Hiding The degree to which a paint obscures the surface to which it is applied.

High-solids systems A system that uses a high-volume, low-pressure spray gun.

High-strength low alloy A type of steel used in unibody design manufacture.

High-strength steel A low-alloy steel that is stronger than hot- or cold-rolled steel; used in the manufacture of structure parts.

High-volume, low-pressure gun A spray gun that atomizes paint into low-speed particles.

Hinge pillar The framework or inner construction to which a door hinge is fastened.

Hold out The ability of a surface to keep the topcoat from sinking in.

Hooding To apply a cover or hood to the headlight area that is constructed of epoxy resin and fiberglass reinforcement.

Hood panel A large metal panel that fills the space between the two front fenders and closes off the engine compartment.

Hot gas welding The use of air or inert gas that is heated by a torch to melt and fuse thermoplastics and plastic filler rods together.

Hot glue gun An electrically heated device that is used to melt and apply adhesive to plastics and other materials.

Hot knife An electrically heated tool that is used for cutting the polyurethane adhesive that is used to secure windshields.

Hot-melt A polymer adhesive that is applied in its molten state.

Hotspot An unprotected area that may be subject to corrosion.

Hot spray A technique of applying hot paint.

HSLA An abbreviation for high-strength low-alloy steel.

HSS An abbreviation for high-strength steel.

Hue A visual characteristic by which one color will differ from another, such as red, blue, and green.

Humidity The amount of water vapor in the air.

Hydraulic The use of a fluid under pressure to do work.

Hydraulic-electric convertible top A system that uses hydraulic cylinders and an electric motor to raise and lower the top.

Hydraulic press A press having a hydraulic jack, or cylinder, that is used for pressing, straightening, assembling, or disassembling components.

Hydraulic tool A tool having a pump system that forces hydraulic fluid into a cylinder to push or pull a ram.

Hydrocarbon A compound that contains carbon and hydrogen.

Hydrometer An instrument used to measure the specific gravity of fluids, such as battery electrolyte.

I-CAR An acronym for Inter-Industry Conference on Auto Collision Repair.

Impact chisel An electrically or pneumatically driven hand tool that creates a hammering and reciprocating action on a chisel bit used to cut metal.

Impact tool An electric or pneumatic driven hand tool used to tighten or loosen bolts and nuts.

Impact wrench An electrically or pneumatically driven hand tool that is used to tighten or loosen nuts and bolts.

Impurities Foreign material, such as paint, rust, or other contaminants, that can substantially weaken a weld joint.

Included angle An angle that places the turning point of the wheel at the center of the tire-to-road contact area.

Independent front suspension A conventional front suspension system in which each front wheel moves independently of the other.

Independent garage A term often used for an independent shop.

Independent rear suspension A rear suspension system that has no cross axle shaft and each wheel acts independently.

Independent shop A repair shop that may be a sole proprietorship or a partnership.

Indicator lamp (1) A bulb used to warn of problems with oil pressure, engine temperature, fuel level, or alternator output. (2) A bulb used for turn signals or hazard signals.

Indirect damage Any damage that occurs away from the point of impact.

Induction heating The generation of heat in a substrate by the application of an electromagnetic field.

Infrared baking The drying of a paint film using heat developed by an infrared source.

Infrared dryer An electrical heating element that emits radiant energy for the drying or curing of automotive finishes.

Infrared light A portion of the spectrum that accounts for most of the heating effects of the sun's rays.

Inhibitor An additive for paint that slows gelling, skinning, or yellowing processes.

Inner panel An automotive body component that adds strength and rigidity to the outer panels.

Install To attach or insert a part, component, or assembly to a vehicle.

Instructor A professional who teaches others automotive mechanics, automotive body repair and refinishing, or any other trade.

Insulation A material that is commonly applied to muffle excessive noise, reduce vibration, and prevent unwanted heat transfer.

Insulator Any material that opposes the flow of electricity.

Insurance adjuster One who reviews estimates to determine which best reflects how the vehicle should be repaired.

Insurance adjustment An agreement between the vehicle owner, insurance company, and the body shop regarding what repairs will be made and who will pay for them.

Integrally welded A term that describes two or more parts that are welded together to form one integral unit.

Interchangeability The ability of new or used replacement parts to fit as well as the original manufactured part.

Interior door trim assembly The coverings and hardware in the cabin surface of a door.

Interior lock A mechanical lever, knob, or button attached via a rod to the lock portion of the latch assembly.

Interior trim All of the upholstery and moldings on the inside of the vehicle.

Internal-mix gun A spray gun that mixes air and material inside the cap.

Internal rust-out Rust damage caused by oxidization from inside to the outside.

Iron A basic element.

Isocyanate resin The principal ingredient in urethane hardeners.

Isopropyl alcohol A solvent that will dissolve grease, oil, and wax, but will not harm paint finishes or plastic surfaces.

Jack A device used for heavy lifting, such as a vehicle.

Jack stand A safety device used to support the vehicle when working under it.

Jig A mechanical device used for positioning and holding work.

Joint The area at which two or more pieces are connected.

Jounce The compression of a spring caused by an upward movement of the wheel and/or a downward movement of the frame.

Jumper wire An electrical test component used to connect or bypass a component for testing.

Jump start The act of connecting a vehicle with a dead battery to a good battery, so that enough current will flow to start the engine.

Jump suit A garment that resists paint absorption, provides full-body protection, and can be worn over other clothing.

Kerf The space left after metal has been removed by cutting with a saw or torch.

Keyless lock system A lock system that operates the lock through use of a numeric keypad on which a code is entered, or by a signal from a small transmitter.

Kick-out The precipitation of the dissolved binder from a solution as a result of solvent incompatibility.

Kick over A term used by some to indicate that a plastic filler has hardened.

Kick pad The panel that fits between the cowl and the front door opening.

Kick panel The panel that fits between the cowl and the front door opening.

Kilogram A unit of measure in the SI metric system.

Kilometer A unit of measure in the SI metric system.

Kink A bend of more than 90° in a distance of less than 0.118 inches (3 mm).

Kinking A method of cold shrinking by using a pick hammer and a dolly to create a series of pleats in the bulged area.

Lace painting A stencil painting technique in which paint is sprayed over fabric lace designs that have been stretched across the surface to be decorated.

Lacquer A type of paint that dries by solvent evaporation and can be rubbed to improve appearance.

Ladder frame A frame design in which the rails are nearly straight with the cross members to stiffen the structure.

Laminar composite Several layers of reinforcing materials that are bonded together with a resin matrix.

Laminated glass Any glass having a plastic film sandwiched between layers for safety.

Lamination A process in which layers of materials are bonded together.

Lap weld A weld that is made along the edge of an overlapping piece.

Laser system A type of measuring system that uses laser optics.

Latch A mechanism that grasps and holds doors, hoods, and trunk lids closed.

Latch assembly A manual- or power-operated handle mechanism for the trunk or hatch.

Late-model A vehicle manufactured within the past fifteen years.

Leader hose A short length of air hose with quick coupler connections used to connect pneumatic tools to the shop air supply.

Leading The act of applying a lead-based solder body filler.

Liability A legal responsibility for business decisions and actions.

Liability insurance Insurance that covers the policyholder on liability damages to the personal property of others.

Lift A hydraulic mechanism used to raise a vehicle off the floor.

Lift channel A channel in which window glass is supported as it is raised and lowered.

Lifting The attack of an undercoat by the solvent in a top coat, resulting in distortion or wrinkling of the undercoat.

Lightbulb An electrical device with internal elements that glow when electrical energy is applied.

Lightness The whiteness of paint measured by the amount of light reflected off its surface.

Liquid sandpaper A chemical that cleans and etches the paint surface.

Liquid vinyl (1) A paint that consists of vinyl in an organic solvent. (2) Material sometimes used to repair holes or tears in vinyl upholstery or similar applications.

Liquid vinyl patching compound A thick material that may be used to repair severely damaged areas of vinyl.

Listing A small pocket in the headlining that holds support rods.

Liter A unit of measure in the SI metric system.

Load Any device or component that uses electrical energy.

Load tester An instrument used to determine the condition of a battery.

Lock A mechanism that prevents a latch assembly from operating.

Lock cylinder The mechanism operated with a key in a mechanical lock system.

Locking cord A term often used for locking strip.

Locking strip The strip that fits into the gasket groove to secure the glass in the gasket.

Lock-out tool A device used to open a door if a door latch is inoperative or the keys are lost or locked inside.

Lock pillar The vertical doorpost containing the lock striker.

Longitudinal A term generally used to identify an engine that is positioned so the crankshaft is perpendicular to the vehicle's axles.

Lord Fusor The trade name for a body panel repair adhesive recommended for bonding panels to space frames.

Low crown A damage area with a slightly convex curve.

Lower glass support A support placed at the bottom edge of the window opening.

Low spot A small concave dent.

Machine guard A safety device used to prevent one from coming into contact with the moving parts of a machine.

Machine rubbing compound A compound with very fine abrasive particles, suitable for machine application.

MacPherson strut A type of independent suspension that includes a coil spring and a shock absorber.

Maintenance-free battery A battery designed to operate its full service life without requiring additional electrolytes.

Major damage Any damage that includes severely bent body panels and damaged frame or underbody components.

Malleability The property that permits metal formation and deformation under compression.

Manager/Supervisor One who has control of shop operations and the hiring, training, promotion, and firing of personnel.

Manual convertible top A top system that is raised and lowered by hand.

Manually operated seat A seat system that is manually adjusted back and forth on a track.

Manual welding A welding process in which the procedure is performed and controlled by hand.

Markup The amount of profit that is added to the cost to determine selling price.

Marred A part or component that has been damaged or scratched.

Mash A vehicle body damage in which the length of any section or frame member is less than factory specifications.

Masking Paper or plastic used to protect surfaces and parts from paint overspray.

Masking paper A special paper that will not permit paint to bleed through.

Masking tape An adhesive-coated paper-back tape used to protect parts from spray paint.

Mass tone The color of paint as it appears in the can or on the painted panel.

Material safety data sheet (MSDS) Data that are available from all product manufacturers detailing chemical composition and precautionary information for all products that can present a health or safety hazard.

Matting Glass-fiber materials loosely held together and used with a resin to make repairs.

Measurement system A system that allows one to check for frame or body alignment or misalignment.

Mechanical fastener A device, such as screws, nuts, bolts, rivets, and spring clips, for the adjustment and replacement of assemblies or components.

Mechanical joining The technique for holding components together through the use of fasteners, folded metal joints, or other means.

Mechanical measuring system A system having a precision beam and tram-like adjustable pointers to verify dimensions.

Mechanism The working parts of an assembly.

Metal conditioner A chemical cleaner used to remove rust and corrosion from bare metal that helps prevent further rusting.

Metal inert gas (MIG) welding A welding technique that uses an inert gas to shield the arc and filler electrode from atmospheric oxygen.

Metal insert A component used to help strengthen and secure the repair when sectioning rocker panels.

Metallic Paint finishes that include metal flakes in addition to pigment.

Metallic paint finish Paint that contains metallic flakes in addition to pigment.

Metallurgy The study of metals and the technology of metals.

Metal snips A hand-held scissor-like tool used to cut thin metal.

Metamerism Two or more colors that match when viewed under one light source, but do not match when viewed under a second light source.

Meter (1) An instrument used to make measurements. (2) A unit of measure in the SI metric system.

Mica A color pigment or particle found in pearl paints.

MIG An acronym for metal arc welding.

MIG spot welding A technique often used to tack panels in place before welding with continuous MIG welds or compression resistance spot welding.

Mil A measure of paint film thickness equal to one one-thousandth of an inch.

Mildew A fungus growth that appears in warm, humid areas.

Mill-and-drill pad An attachment point used when sectioning a plastic body panel.

Millimeter A unit of measurement in the SI metric system.

Mineral spirits A petroleum-based product having about the same evaporation rate as gum turpentine; sometimes used for wet sanding and to clean spray guns.

Minor damage Any repair that requires relatively little time and skill to complete.

Mirror bracket adhesive A strong bonding material used to mount a rear-view mirror on the inside of a windshield.

Misaligned Uneven spacing, as between body panels.

Miscible Capable of being mixed.

Mist Liquid droplets suspended in the air due to condensation from the vapor to liquid state, or by breaking up a liquid into a dispersed state by atomizing.

Mist coat A light spray coat of high-volume solvent for blending and/or gloss enhancement.

Model year The production period for new model vehicles or engines.

Mold core method A procedure used to repair curved or irregularly shaped sections.

Molding clip A mechanical fastener used to secure trim.

Molecule The smallest possible unit of any substance that retains characteristics of that substance.

Monocoque A unibody vehicle construction type in which the sheet metal of the body provides most of the structural strength of the vehicle.

Motorized seat belt A seat belt system designed to automatically apply the shoulder belts to the front-seat passengers.

Mottling A paint film defect that appears as blotches or surface imperfections.

Movable glass A window designed to be moved up and down or side-to-side for opening and closing.

Movable section A component held in position by a mechanical fastener.

Mud A slang term used for ready-to-use plastic filler.

Multimeter An electrical instrument that is used to measure resistance, voltage, and amperage.

Multiple-pull system A system that pulls in two or more directions to correct damage.

Music wire Steel wire used for cutting through adhesives.

Negative caster A condition occurring when the top of the steering knuckle is tilted toward the front of the vehicle.

Negligent To be careless or irresponsible.

Net The amount of money left after paying all overhead expenses; known as profit.

Neutral flame The flame of an oxyacetylene torch that has been adjusted to eliminate all of the inner cone acetylene feather.

Neutralizer Any material used to chemically remove any trace of paint remover before finishing begins.

Nibbler A power hand tool used to cut small bites out of sheet metal.

Nitrile glove Gloves used for protection when working with paints, solvents, catalysts, and fillers.

No-fault insurance A type of insurance that covers only the insured's vehicle and/or personal injury, regardless of who caused the accident.

Noise intensity The loudness of a noise.

Non-bleeder gun A paint spray gun having a valve that shuts off the air flow when the trigger is released.

Noncompetitive estimate A detailed and accurate estimate usually done for minor damage where no claims are to be filed with an insurance company.

Nonferrous metal A metal that contains no iron, such as aluminum, brass, bronze, copper, lead, nickel, and titanium.

Nose The front body section ahead of the doors, including bumper, fenders, hood, grille, radiator, and radiator support.

OEM An abbreviation for original equipment manufacturer.

Office staff Those who perform office duties such as billing, receiving payments, making deposits, ordering parts, and paying bills.

Off-the-dolly dinging A technique of holding the dolly away from the raised areas being hit by the hammer.

Ohm A unit of measure for resistance.

Ohmmeter An electrical device used to measure resistance.

Oil A viscous liquid lubrication product derived from various natural sources, such as vegetable oil.

One-wire system The electrical wiring system for a vehicle that uses the chassis as an electrical path to ground, eliminating the need for a second wire.

On-the-dolly dinging A technique of holding the dolly directly under the area where the hammer is used.

On-the-job training A method whereby the beginning technician learns the trade from experienced technicians while taking part in hands-on repairs.

Opaque Not transparent, or impervious to light.

Open bid The scenario in an estimate whereby a part may be suspect of needed repair or replacement but cannot be determined until the repairs are under way.

Open circuit An incomplete circuit due to a break or other interruption that stops the flow of current.

Open-coat abrasive An abrasive in which the abrasive grains are widely separated.

Open structural member A flat panel accessible from either side, such as a floor panel.

Orange peel An irregularity in the surface of a paint film that appears as an uneven or dimpled surface but feels smooth to the touch.

Orbital sander A hand-held power sanding tool that operates in an elliptical or oval pattern.

Orifice A small calibrated opening.

Original finish The paint that is applied by the vehicle manufacturer.

Outer-belt weather strip The material that is located between the door panel and the window glass to prevent dirt, air, and moisture from entering.

Outer panel The sheet metal section that is attached to an inner panel to form the exterior of a vehicle.

Outside surface rust Rust that starts on the outside of a panel.

Oven Equipment that is used to bake on a finish.

Overage The added charges for any damage that may be discovered after the original estimate.

Overall repainting Refinish repair that includes completely repainting the vehicle.

Overhaul A procedure in which an assembly is removed, cleaned, and/or inspected, and damaged parts are replaced, rebuilt, and reinstalled.

Overhead dome light Lights that provide illumination in the interior of a vehicle.

Overlap (1) The spray that covers the previous spray stroke. (2) In estimating, when two operations share common steps or procedures, thereby the same flat-rate charges.

Overlay A thin layer of decorative plastic material often with a design or pattern applied to various parts of the vehicle.

Pitting The appearance of holes or pits in a paint film while it is wet.

Plasma A gas that is heated to a partially ionized condition enabling it to conduct an electric current.

Plasma arc cutting A cutting process in which metal is severed by melting a localized area with an arc and removing the molten material with a high velocity jet of hot, ionized gas.

Plastic A manufactured lightweight material that is now being used in automobile construction.

Plastic alloy A material that is formed when two or more different polymers are mixed together.

Plastic deformation The use of compressive or tensile force to change the shape of metal.

Plastic filler A compound of resin and fiberglass used to fill dents and level surfaces.

Plasticity The property that allows metal to be shaped.

Plasticizer A material that is added to paint to make film more flexible.

Plastic-resin mixture A material used to fill chips and pits on a windshield.

Platform frame A frame that consists of a floor pan and a central tunnel.

Pliers A hand tool designed for gripping.

Pliogrip A body repair adhesive used for bonding panels to space frames.

Plug weld The adding of metal to a hole to fuse all metal.

Pneumatic flange tool An air operated hand tool used to form an offset crimp along the joint edge of a panel.

Pneumatic tool A tool that is powered by compressed air.

Polisher A term used for buffer.

Polishing compound A fine abrasive paste used for smoothing and polishing a finish.

Polishing cream An extremely fine abrasive material used for manual removal of swirl marks left after machine compounding.

Polyblend A plastic that has been modified by the addition of an elastomer.

Polyester resin A thermosetting plastic used as a finish and matrix binder with reinforcing materials.

Polyethylene A thermoplastic used for interior applications.

Polypropylene A thermoplastic material used for interior and some underhood applications.

Polyurethane A chemical compound that is used in the production of resins for enamels.

Polyurethane adhesive A plastic compound used with butyl tape to bond fixed glass in place.

Polyurethane enamel A refinishing material that provides a hard, tile-like finish.

Polyurethane foam A plastic material used to fill pillars and other cavities, adding strength, rigidity, and sound insulation.

Polyurethane primer A material that may be brushed onto the areas where adhesive will be applied to hold glass in place.

Polyvinyl chloride A thermoplastic material used in pipes, fabrics, and other upholstery materials.

Pool An area in molten metal that is created by the heat of the welding process.

Poor drying A condition in which a finish stays soft and does not dry or cure as quickly as the painter may like.

Pop rivet gun A hand-held tool designed to place and secure rivets into a blind hole.

Porosity Voids or gas pockets in any material.

Portable alignment system An alignment system used for correcting frame and body damage.

Portable grinder A hand-held tool used for grinding.

Positive caster The condition that results when the top of the steering knuckle is tilted toward the rear of the car.

Positive post The positive terminal of a battery.

Pot life The time a painter has to apply a plastic or paint finish to which a catalyst or hardener has been added before it will harden.

Pot system A rail alignment system that uses portable hydraulic units anchored to attachment holes in the floor.

Power ram A hydraulic body jack used to correct severe damage.

Power ratchet An electrical or pneumatic powered tool used to remove and replace nuts, bolts, and other fasteners.

Power seat A seat that may be adjusted vertically and horizontally with the use of electric motors.

Power source A source of electrical energy, such as the battery.

Power tool A tool that operates off electrical, hydraulic, or pneumatic power.

Power washer A cleaning machine that uses a high-pressure spray of water to dislodge debris.

PPM An abbreviation for parts per million.

Press fit A joining technique in which one part is forced into the other.

Pressure A force measured in pounds per square inch (psi) or kiloPascals (kPa), such as the air delivered to a paint spray gun.

Pressure drop A loss of air pressure between the source and the point of use.

Pressure feed gun A paint spray gun in which air pressure or a high-pressure pump force the paint to the gun.

Pressure pot A paint spray system in which paint is fed to the spray gun by air pressure.

Pressurize To apply a pressure that is greater than atmospheric pressure.

Primary damage The damage that occurs at the point of impact on a vehicle.

Prime coat The first coat to improve adhesion and provide corrosion protection.

Primer A type of paint that is applied to a surface to increase its compatibility with the topcoat and to improve adhesion or corrosion resistance.

Primer-sealer An undercoat that improves the adhesion of a topcoat and seals old painted surfaces that have been sanded.

Primer-surfacer A high-solids sandable primer that fills small voids and imperfections.

Priming A process to smooth the surface and help the paint topcoat to bond.

Proprietorship A business, such as a body shop, that is owned by one person.

Puller (1) A tool used to pull out dents. (2) A tool used to remove hubs and pulleys.

Pulling Applying a force.

Pull rod A tool that allows repairs to be performed from the outside of a damaged panel.

Pull tab A metal tab welded to a damaged panel that allows the use of a slide hammer.

Putty A material that is used to fill flaws.

PVC (1) An abbreviation for polyvinyl chloride, a type of plastic. (2) An abbreviation for pigment volume content, the percentage of pigment in solid material of a paint.

Quarter panel The side panel extending from the rear door to the end of a vehicle.

Quench To cool quickly.

Rack-and-pinion steering A steering gear in which a pinion gear on the end of the steering shaft meshes with a rack gear on the steering linkage.

Rail The major member that forms the box-like support in unibody construction.

Rail system A specially designed steel member set into the shop floor providing an anchorage for pushing and pulling equipment.

Reaction injection molding A process involving injecting reactive polyurethane or a similar resin onto a mold.

Rear clip The rear portion of the car including part of the roof.

Rear compartment lid The trunk lid or panel and reinforcement that covers the luggage compartment.

Rear-engine An engine that is positioned directly above or slightly in front of the rear axle.

Rechrome To replace a part, such as a bumper, with chrome.

Reciprocating sander A hand-held power sander with a sanding surface that moves in small circles while moving in a straight line.

Reduce To lower the viscosity of a paint by the addition of solvent or thinner.

Reducer A solvent combination used to thin enamel.

Reference point The point on a vehicle, including holes, flats, or other identifying areas, used to position parts during repairs.

Refinish To repaint by removing or sealing an old finish and applying a new topcoat.

Reflow A process by which lacquers are melted to produce better flow characteristics.

Regulator (1) A mechanism used to raise or lower glass in a vehicle door. (2) A device used to control pressure of liquids or gases.

Reinforced reaction injection molding A process that involves injecting reactive polyurethane, polyurea, or dicyclopentadiene resin into a mold that contains a preformed glass mat.

Reinforcement piece A sheet metal welded in place along a joint to strengthen the joint.

Reinforcements Structural braces used to strengthen panels.

Relative density The mass of a given volume compared to the same volume of water at the same temperature, referred to as specific gravity.

Relief valve A safety valve designed to open at a specified pressure.

Remove and reinstall A term for removing an item to gain access to a part, then reinstalling the item.

Remove and repair A term for removing and reinstalling a part.

Replacement panel A body panel used to replace a damaged panel.

Resin A term used for polyester or epoxy resin.

Resistance An opposition to the flow of electricity.

Resistance weld A weld made by passing an electric current through metal between the electrodes of a welding gun.

Resource Conservation and Recovery Act A law passed to enable the Environmental Protection Agency to control, regulate, and manage hazardous waste generators.

Respirator A mask worn over the mouth and nose to filter out particles and fumes from the air being breathed.

Retaining clip A spring-type device used to secure one component to another.

Retaining strip A strip sewn into the head liner and attached to T-slots on the inner roof panel to support the head liner.

Retarder A slow evaporating additive used to slow drying.

Reveal file A curved file used to shape tight curves or rounded panels.

Reveal molding An exterior trim piece used to accent a glass opening.

Reverse masking A spot repair technique used to help blend the paint and make the repair less noticeable.

Reverse polarity The connecting of a MIG welding gun to the positive terminal of the DC welder providing greater penetration.

Right-to-know law One's right to essential information and stipulations for safely working with hazardous materials.

Rocker panel A narrow panel attached below the car door that fits at the bottom of the door opening.

Roof rail The framework or inner construction that reinforces and supports the sides of the roof panel.

Roughing out The preliminary work of bringing the damaged sheet metal back to the approximate original contour.

Rubber stop An adjustable hood bumper used to make minor alignment adjustments.

Rubbing compound An abrasive that smooths and polishes paint film.

Runs and sags A heavy application of sprayed material that fails to adhere uniformly to the surface.

Rust Corrosion that forms on iron or steel when exposed to air and moisture.

Rust inhibitor A chemical applied to steel to retard rusting or oxidation.

Rust-out A condition that occurs when rust is allowed to erode through a metal panel.

SAE An abbreviation for the Society of Automotive Engineers.

Safety glasses Protective eyewear.

Safety shoes Protective footwear.

Safety stand A metal support that is placed under a raised vehicle.

Sag Frame damage in which one or both side rails are bent and sag at the cowl.

Sagging Excessive paint flow on a vertical surface resulting in drips and other imperfections.

Salary A fixed dollar amount that is paid a worker per day, week, month, or year.

Sales personnel A manufacturer or equipment suppliers representative who sells various products or services.

Salvage The value of a wrecked vehicle that has been declared beyond repair.

Sandblasting A method of cleaning metal using an abrasive, such as sand, under air pressure.

Sander A hand-held power tool used to speed the rate of sanding or polishing surfaces.

Sanding The use of an abrasive coated paper or plastic backing to level and smooth a body surface being repaired.

Sanding block A hard flexible block that provides a smooth backing for hand sanding operations.

Sand scratches Marks that are made in metal or an old finish by an abrasive.

Sandscratch swelling A swelling of sand scratches in a surface caused by solvents in the topcoat.

Saturate To fill an absorbent material with a liquid.

Scanner An electronic device for reading and storing data or computer information.

Scraper A hand-held tool used to scrape away paint or other surface material.

Scratch awl A pointed tool used for marking and piercing sheet metal.

Screwdriver A hand-held, pneumatically or electrically powered tool used to tighten or loosen screw-type fasteners.

Scuff To roughen a surface by rubbing lightly with sandpaper to provide a suitable surface for painting.

Sealed beam headlight A light in which the filament, reflector, and lens are fused into a single hermetic unit.

Sealer A coat between the topcoat and the primer or old finish to give better adhesion.

Sealing strip A strip inside the door that prevents dust from entering the drain holes while allowing water to drain.

Seat belt A restraint that holds occupants in their seats.

Secondary damage The indirect damages that may occur due to misplaced energy that causes stresses at areas other than the primary impact zone.

Sectioning The act of replacing partial areas of a vehicle.

Seeding The development of small insoluble particles in a container of paint which results in a rough or gritty film.

Self-centering gauge A device used to show misalignment.

Self-contained respirator A compressed air cylinder equipped respirator that provides protection.

Self-etching primer A primer that contains an agent that improves adhesion.

Semigloss A gloss level between high gloss and low gloss.

Series circuit An electrical circuit with one or more loads wired so the current has only one path to follow.

Service manual A book published by the vehicle manufacturer that lists specifications and service procedures.

Setting time The time it takes solvents to evaporate or resins to cure or become firm.

Settling The separation caused by gravity of one or more components from a paint that results in a layer of material at the bottom of a container.

Shaded glass Glass having a dark color band across its top portion.

Shading A custom painting technique accomplished by holding a mask or card in place and overspraying the surrounding area.

Sheen The gloss or flatness of a film when viewed at an angle.

Sheet molding compound A thermoset composite that can be formed into strong and stiff body components.

Shelf life The length of time the manufacturer recommends that a material may be stored and remain suitable for use.

Shim A thin metal piece used behind panels to bring them into alignment.

Shock tower The reinforced body areas for holding the upper parts of the suspension system.

Shop layout The arrangement of work areas, storage, aisles, office, and other spaces in a shop.

Shop manual A term often used for service manual.

Shop tool A major tool that the shop owner usually provides.

Short circuit An electrical leakage between two conductors or to ground.

Short method A partial replacement of adhesive when installing new fixed glass.

Show through Sand scratches in an undercoat that are visible through the topcoat.

Shrinking The act of removing a bulge from metal by hammering with a hammer and dolly, with or without heat.

Shrinking dolly A hand tool used with a shrinking hammer to reduce the surface area of metal without using heat.

Shrinking hammer A hand tool used with a shrinking dolly to reduce the surface area of metal without using heat.

Side sway Damage that occurs when an impact to the side of a vehicle causes the frame to bend or wrinkle.

Silicon carbide An abrasive used in sanding or grinding grit.

Silicone An ingredient in waxes and polishes which makes them feel smooth.

Silicone adhesive An adhesive used to repair torn vinyl trim and upholstery.

Silicone-treated graining paper Special paper that is used to create the final grain pattern in vinyl during upholstery repair.

Simulated seam tape Tape that is used to provide the appearance of seams in a spray-on vinyl roof covering.

Single coat Passing one time over the surface with each stroke overlapping the previous coat by 50 percent.

Single pull system A straightening system capable of only pulling in one direction at a time.

Siphon feed gun A type of paint spray gun in which the paint is drawn out of the container by vacuum action.

Skin The outer panel.

Skinning The formation of a thin tough film on the surface of a liquid paint film.

Slide caliper rule A rule used to measure inside or outside dimensions and the depth of a hole.

Slide hammer A hand-held tool having a hammer head that is slid along a rod and against a stop so that it pulls against the object to which the rod has been fastened.

Snap fit A joining technique in which the parts are forced over a lip or into an undercut retaining ring.

Sodium hydroxide powder A powder that is produced when the sodium azide pellets in an air bag are activated.

Soft-faced hammer A hammer having a head of plastic, wood, or other soft material.

Software Computer information stored in floppy disks, computer programs, and CDs.

Solder A mixture of tin, lead, antimony, or silver, that may be melted to fill dents and cracks in metal or to join wires in an electrical circuit.

Soldering A joining process in which the base metal is heated enough to allow the solder to melt and make an adhesive bond.

Solder paddle A wooden spatula-type tool used for applying and spreading body solder.

Solids The percentage of solid material in paint after solvents have evaporated.

Solvency The ability of a liquid to dissolve resin or any other material.

Solvent A liquid which will dissolve something, such as plastic.

Solvent cement A thin liquid that partly dissolves plastic materials so they may be bonded together.

Solvent popping The blisters that form on a paint film that are caused by trapped solvents.

Sound deadening material A pad or sheet of plastic material that absorbs sound.

Space frame A variation of unitized body construction in which molded plastic panels are bonded with adhesive or mechanical fasteners to a space frame.

Spanner socket A specialty tool used for special applications, such as the removal of antennas, mirrors, and radio trim nuts.

Specialty shops A shop that specializes in a certain area, such as frame straightening, wheel alignment, upholstering, or custom painting operations.

Specifications Data supplied by the manufacturer covering all measurements and quantities of the vehicle.

Specific gravity The ratio of the weight of a specific volume of a substance in the air compared to the weight of an equal volume of water.

Spectrophotometer An instrument to measure color.

Speed file A long holder used with strips of coarse sandpaper to smooth a work area.

Spider webbing A custom painting effect produced by forcing acrylic lacquer from the spray gun in the form of fibrous thread.

Spitting A paint spray gun problem caused by dried-out packing around the fluid needle valve.

Spontaneous combustion A process of material igniting by itself.

Spoon A tool used in the same manner as a dolly designed for use in confined areas and to pry panels back into position.

Spot cutter bit A tool used to cut through the welds on a panel.

Spot putty A plastic-like material used for filling small holes or sand scratches.

Spot repair A type of repair in which a section of a car smaller than a panel is repaired and refinished.

Spot weld A weld in which an arc is directed to penetrate both pieces of metal.

Spray gun A hand-held painting tool powered by air pressure that atomizes liquids, such as paint.

Spray mask A thin film that is sprayed on the surface to be decorated so a design is cut through the film and the desired portions may be removed before paint is applied.

Spray-on roof covering A vinyl coating sprayed in place.

Spray pattern A cross section of the spray.

Spreader A rigid rectangular piece of plastic used to spread body filler.

Spread ram A tool having two jaws that are forced apart when the tool is activated.

Squeegee A flexible rubber-like tool used to apply body putty or filler.

Stationary rack system A system that can be used to repair extensive collision damage.

Stationary section The permanent assemblies or components of a vehicle that cannot be moved.

Steel A ferrous metal used in the construction of a vehicle and as a substrate for paint, which must be painted to prevent corrosion.

Steering alignment specialist A technician who specializes in alignment and wheel balancing, as well as repairing steering mechanisms and suspension systems.

Steering axis inclination The inward tilt of the steering knuckle.

Steering system The mechanism that enables the driver to turn the wheels to change the direction of a vehicle's movement.

Stencil An impervious material into which designs have been cut.

Stitch welding The use of intermittent welds to join two or more parts.

Stools A low seat equipped with wheels.

Stop A check block or spacer used in movable glass installations.

Storage battery The device that converts chemical energy into electrical energy.

Straight-in damage The damage that results from a direct impact.

Strainer A fine mesh screen that is used to remove small lumps of dirt or other debris from a liquid.

Strength (1) The measure of the ability of a pigment to hide color. (2) The integrity of a structure.

Stress To relieve or take tension off a part.

Stress line The low area in a damaged panel that usually starts at the point of impact and travels outward.

Stretching The deformation of metal under tension.

Striker pin A bolt that can be adjusted laterally, vertically, fore, and aft to achieve door clearance and alignment.

Striker plate That portion of the door lock that is mounted on the body pillar.

Stringer bead A weld made by moving the electrode in a straight continuous line.

Striping brush A small brush used to apply stripes and other designs by hand.

Striping tool A tool with a small paint container and a brass wheel that applies paint as the tool is moved along the surface.

Stripper A term used for paint remover.

Stripping The act of removing paint by applying a chemical which softens and lifts it, by using air-powered blasting equipment, or by power sanding.

Structural adhesive A strong flexible thermosetting adhesive.

Structural integrity The body strength and ability of a vehicle to remain in one piece.

Structural member A load-bearing portion of the body structure that affects its over-the-road performance or crash worthiness.

Structural panel A panel used in a unibody that becomes a part of the whole unit and is vital to the strength of the body.

Strut suspension A suspension system that attaches to the spring tower and lower control arm.

Stub frame A unibody with no center rail portions but with front and rear stub sections.

Stud A headless bolt having threads on one or both ends.

Stud welding The joining of a metal stud or similar part to a work piece.

Subassembly An assembly of several parts that are put together before the whole is attached.

Sub-frame A unitized body frame only having the front and rear stub sections of frame rails.

Sublet Repair or services sent out to another shop.

Submember A box- or channel-like reinforcement that is welded to a vehicle floor.

Substrate The surface that is to be worked on.

Suction cup A rubber or plastic cup-like device that is used to hold and position large sections of glass.

Sunroof A vehicle roof having a panel that slides back and forth on guide tracks.

Sunroof panel A transparent section of the roof that can be removed or slid into a recess for light or ventilation.

Support rod A metal rod that supports the head liner in a vehicle.

Surface preparation To prepare an old surface for refinishing or painting.

Surfacer A heavily pigmented paint that is applied to a substrate to smooth the surface for subsequent coats of paint.

Surface rust Rust found on the outside of a panel that has not penetrated the steel.

Surface-scratching method The scratching of an arc welding electrode across the work to form the starting arc.

Surfacing mat A thin fiberglass mat used as the outer layer when making repairs.

Surform A surface forming grater file with open, rasp-like teeth.

Suspension system The springs and other components supporting the upper part of a vehicle on its axles and wheels.

Swirl remover A term used for polishing cream.

Symmetrical design A design in which both sides of a unibody are identical in structure and measurement.

Syntactic foam A resin and catalyst system that contains glass or plastic spheres used to fill rusted areas in door sills and rocker panels.

Tack The stickiness of a paint film or adhesive.

Tack cloth Cheesecloth that has been treated with nondrying varnish to make it tacky to pick up dust and lint.

Tack coat A light dusting coat that is allowed to become tacky before applying the next coat.

Tack rag A resin-impregnated cheesecloth used to pick up small dust and lint from a surface before being painted.

Tack weld A temporary weld to hold parts in place during final welding operations.

Tap A device used to cut internal threads.

Tap and die The tools used to restore threads or to cut new threads.

Tape measure A retractable measuring tool.

Tapping technique The momentary touching of an arc welding electrode to the work as a means of striking an arc.

Technical pen A pen used to draw pinstripes by hand on a vehicle.

Temperature-indicating crayon A temperature sensitive crayon used to mark across a weld area to monitor its temperature.

Temperature make-up system A heating/cooling system that filters and conditions the air before it enters a paint spray booth.

Tempered glass Glass that has been heat-treated.

Tensile strength The resistance to distortion.

Terminal A mechanical fastener attached to wire ends.

Test light An electrical test device that will light when voltage is present in a circuit.

Texturing agent The material that is added to paint to produce a bumpy texture.

Thermoplastic A plastic material that softens when heated and hardens when cooled.

Thermosetting A solid that will not soften when it is heated.

Thermosetting plastic A resin that does not melt when it is heated.

Thinner A solvent used to thin lacquers and acrylics.

Three-stage Three paint layers that produce a pearlescent appearance consisting of a basecoat, a midcoat, and a clearcoat.

Thrust line An imaginary line parallel with the rear wheels.

TIG An abbreviation for tungsten inert gas arc welding.

Tinning The melting of a solder and flux onto an area to be soldered.

Tint (1) A light color, usually pastel. (2) To add color to white or another color.

Tinted glass Window glass having a color tint.

Tinting strength The ability of a pigment to change the color of a paint to which it is added.

Tint tone The shade that results when a color is mixed with white paint.

Toe The position of the front of a wheel when compared to the rear of the wheel.

Toe in A condition whereby the front edge of the wheels are closer together than the rear edge of the wheels.

Toe out A condition whereby the rear edge of the wheels are closer together than the front edge of the wheels.

Tolerance The acceptable variation, plus or minus, of vehicle dimensions as provided by the manufacturer.

Topcoat The top layer of paint applied to a substrate.

Torque box A structural component provided to permit some twisting as a means of absorbing road and collision impact shock.

Torsion bar A metal bar that is twisted as a lid is closed to provide a spiral tension.

Total loss A situation whereby the cost of repairs would exceed the vehicle value.

Touch-up gun A paint spray gun, similar to a conventional spray gun, but with a smaller capacity used for touch-up, stenciling, and small detail.

Touch-up paint A small container of paint matched to the factory color used to fill small chips in a vehicle finish.

Tower The upright portion of a frame-straightening system.

Tower cut The same as a nose, but includes the shock tower or strut tower.

Toxic fumes Harmful fumes that can cause illness or death.

Toxicity The biological property of a material reflecting its inherent capacity to produce injury or an adverse effect due to overexposure.

Tracer A colored coding stripe on an electrical wire for identification when tracing a circuit.

Tracking The ability of the rear wheels of a vehicle to follow the front wheels.

Tracking gauge A tool used to detect and measure the misalignment of front and rear wheels.

Track molding A trim piece consisting of a metal track and a plastic insert.

Tram gauge An instrument used to check alignment and dimensions against factory specifications.

Transmitter/receiver lock system A system whereby a signal is transmitted from a small key ring transmitter to the receiver located in the door to activate the locking mechanism.

Transverse An engine positioned so that its crankshaft is parallel to the vehicle's axles.

Trim A decorative metal piece on a vehicle body.

Trim cement An adhesive used to attach upholstery and selected trim.

Tunnel A formation in the floor panel for transmission and drive shaft clearance on a rear-wheel-drive vehicle.

Turning radius The amount one front wheel turns sharper than the other.

Turpentine A solvent derived from the distillate of pine trees.

Twist Collision damage that causes distortion of the frame cross members.

Two-component epoxy primer A primer material having two components that react after being mixed together.

Two-part A product supplied in two parts which must be mixed together in correct proportions immediately before use.

Two-stage Two coats of paint such as a basecoat and a clearcoat.

Two-tone Two different colors on a single paint scheme.

Ultrasonic plastic welding A technique of repairing rigid auto plastics in which welding time is controlled by a power supply.

Ultrasonic stud welding The variation of a shear joint used to join plastic parts whereby a weld is made along a stud's circumference.

Ultraviolet (UV) light That part of the invisible light spectrum which is responsible for degradation of paints.

Ultraviolet (UV) stabilizer A chemical added to paint to absorb ultraviolet radiation.

Underbody The lower portion of a vehicle that contains the floor pan, trunk floor, and structural reinforcements.

Undercoat The first coat of a primer, sealer, or surfacer.

Undercoating (1) The second coat of a three-coat finish; the first coat in repainting. (2) The coating or sealer on the underside of panels to help prevent rust and deaden sound.

Undercut A groove in the base metal adjacent to a weld and left unfilled by weld metal.

Unibody A vehicle style whereby parts of the body structure serve as support for overall vehicle strength.

Unibody construction A vehicle construction type in which the body and underbody form an integral structural unit.

Universal measuring system A measuring system having frame-mounted devices that can be adjusted for various vehicle bodies.

Universal thinner A solvent that is used to thin lacquers and to reduce enamels.

Upholsterer One whose expertise is the repair or replacement of interior surface materials.

Upholstery ring pliers Pliers that are used to remove or install upholstery rings.

Upsetting The deformation of metal under compression.

Up-stop A component that limits the upward travel of the lift channel.

Urethane A type of paint or polymer coating noted for its toughness and abrasion resistance.

Utility knife A hand-held cutting tool having a replaceable retractable blade.

UV stabilizer A chemical added to paint to absorb ultraviolet radiation.

Vacuum Any pressure below atmospheric pressure.

Vacuum cleaner A portable suction device used to clean vehicle interiors.

Vacuum cup puller A large suction cup used as a dent puller.

Vacuum patch A device that is placed over a glass repair area to withdraw all air to ensure that all voids are filled with resin.

Value (1) The darkness or lightness of a color. (2) The fair cost or price of an item.

Vapor A state of matter; a gaseous state.

Vehicle (1) A car or a truck. (2) All of a paint except the pigment including solvents, diluents, resins, gums, and dryers.

Vehicle identification number (VIN) A number that is assigned to vehicles by the manufacturer for registration and identification purposes.

Veil A term used for a surfacing mat.

Veiling The formation of a web or strings in a paint as it emerges from a spray gun.

Ventilation fan An electrical device used to remove vapors and fine particles from the work area.

Vibrating knife A knife with a rapidly moving blade that cuts through polyurethane adhesives.

Vinyl A class of material which can be combined to form vinyl polymers used to make chemical resistant finishes and tough plastic articles.

Vinyl-coated fabrics Any material with a plastic protective or decorative layer bonded to a fabric base that provides strength.

Vinyl paint Any paint material applied to a vinyl top to restore color.

Viscosity (1) The consistency or body of a paint. (2) The thickness or thinness of a liquid.

Visual estimate A guess by an experienced estimator relative to the cost of repairing damage.

VOC An abbreviation for volatile organic compound.

Volatile A material that vaporizes easily.

Volatile organic compound A hydrocarbon that readily evaporates into the air and is extremely flammable.

Volatility The tendency of a liquid to evaporate.

Volt A unit of measure of electrical pressure.

Voltage The electrical pressure that causes current to flow.

Voltmeter An electrical device used to measure voltage.

Wage The amount of money paid to workers, generally computed on a "per hour" basis.

Warpage The distortion of a panel during heat shrinking.

Wash thinner A low-cost solvent used to clean spray guns and other equipment.

Water/air shield A deflector built into a vehicle door.

Waterproof sandpaper Sandpaper that may be used with water for wet sanding.

Water spotting A condition created when water evaporates on a finish before it is thoroughly dry.

Wax (1) A slippery solid sometimes added to paints to add some property. (2) A prepared material used to shine or improve a surface.

Weathering A change in paint film caused by natural forces such as sunlight, rain, dust, and wind.

Weather stripping A rubber-like gasket used to keep dirt, air, and moisture out of the passenger or trunk compartments of a vehicle.

Weave bead A wide weld bead made by moving the electrode back and forth in a weaving motion.

Weld The act of joining two metal or plastic pieces together by bringing them to their melting points.

Welding The joining method that involves melting and fusing two pieces of material together to form a permanent joint.

Weld through primer A primer applied to a joint before it is welded to help prevent galvanic corrosion.

Wet coat A heavy coat of paint.

Wet-on-wet finish A technique of applying a fresh coat of paint over an earlier coat which has been allowed to flash but not cure.

Wet sanding Sanding using a water resistant, ultrafine sandpaper and water to level paint.

Wet spot A discoloration caused where paint fails to dry and adhere uniformly.

Wheel alignment The positioning of suspension and steering components to assure a vehicle's proper handling and maximum tire wear.

Wheel balancing The act of properly distributing weight around a tire and wheel assembly to maintain a true running wheel perpendicular to its rotating axis.

Wheel base The distance between front and rear axles.

Wheelhouse The deep curved panels that form compartments in which the wheels rotate.

Whipping The improper movement of a paint spray gun that wastes energy and material.

Wind cord A rope-like trim placed around doors to help seal and decorate the openings.

Wind lace A rope-like trim placed around doors to help seal and decorate the openings.

Window channel The grooves, guides, or slots in which the glass slides up and down or back and forth.

Window regulator The door-mounted mechanism that provides a means of cranking the window up and down.

Window stop A device inside a door used to limit glass height and depth.

Window tool Tools required to properly remove and install window glass.

Windshield pillar The structural member that attaches the body to the roof panel.

Wiring diagram Drawings that show where wires are routed and how components are arranged.

Wiring harness Electrical wires gathered in a bundle.

Wood grain transfer A plastic transfer sheet used to simulate wood on the sides of a vehicle.

Work hardening Metal that has become stiffer and harder in the stretched areas due to permanent stresses.

Wrench A hand tool used to turn fasteners.

Wrinkle (1) The pattern formed on the surface of a paint film by improperly formulated or specially formulated coatings. (2) The appearance of tiny ridges or folds in paint film.

Wrinkling A surface distortion that occurs in a thick coat of enamel before the underlayer has properly dried.

X-checking The process of taking measurements and comparing them to corresponding dimensions on the opposite side of the vehicle to reveal damage.

X-frame A frame design that does not rely on the floor pan for torsional rigidity.

Zahn cup A paint cup with a hole in the bottom that is used to measure a material's viscosity.

Zebra effect A streaky looking metallic finish, usually caused by uneven application.

Zero plane A plane that divides the datum plane into front, middle, and rear sections.

Zinc A metal coating used to prevent corrosion.

Zinc chromate A paint material that is used for primer to protect steel and aluminum against rust and corrosion.

Zoning A method of systematically observing a damaged vehicle.

Zoning ordinance The law that limits the type of businesses allowed in a particular area or zone.